Costas Loops

Roland Best

Costas Loops

Theory, Design, and Simulation

 Springer

Roland Best
Oberwil
Switzerland

Additional material to this book can be downloaded from
http://www.springerlink.com/978-3-319-72008-1

ISBN 978-3-319-89130-9   ISBN 978-3-319-72008-1   (eBook)
https://doi.org/10.1007/978-3-319-72008-1

Printed on acid-free paper

This Springer imprint is published by Springer Nature
The registered company is Springer International Publishing AG
The registered company address is: Gewerbestrasse 11, 6330 Cham, Switzerland

Preface

The Costas loop has been invented in 1956 by an American engineer J. P. Costas. His original circuit was used for the synchronous demodulation of double-sideband amplitude-modulated signals with suppressed carrier. What Costas designed that time was a variant of the phase-locked loop (PLL), a circuit that had been known long before. We will see later why the conventional PLL failed in such an application. Today, Costas loops are mainly used for the detection of signals by making use of digital modulation techniques, such as binary phase shift keying (BPSK). It showed up that a BPSK signal has very similar properties like the formerly mentioned amplitude-modulated signal. Later, the Costas loop was extended for the application in quadrature phase shift keying (QPSK) and also in m-ary PSK. Costas loops are also found today in the demodulation of quadrature amplitude modulation (QAM) signals.

Like the PLL, the Costas loop is a synchronizing device. The incoming signal in both systems is usually a carrier having frequency f_C that is modulated with the transmitted signal. Both Costas loop and PLL incorporate a local oscillator, operating at frequency f_{Loc}, and this frequency is controlled in such a way that it locks onto the carrier in both frequency and phase, hence the name "phase-locked loop." When a data transmission starts or when a PLL or Costas loop is switched on, the initial frequency f_{Loc} is not yet synchronized to the carrier frequency, but it must first get locked to that frequency. This process is referred to as acquisition process. With the PLL, two different acquisition processes have been defined: (1) the relatively fast lock-in process and (2) the slower pull-in process. For the PLL, a quantity called lock range f_L has been defined. When the initial frequency of the local oscillator is within that lock range, the system will lock within at most one beat note between carrier frequency f_C and initial local oscillator frequency f_{Loc}. The time to get locked is called lock time T_L. When the initial frequency of the local oscillator is outside the lock range but within another range called pull-in range f_P, acquisition will still take place but is much slower. The time required for the pull-in process is called pull-in time T_P. The dynamic performance has been extensively investigated in case of the PLL; here, the designer can make use of equation enabling to compute all these parameters (lock range, lock time, pull-in range, pull-in time) explicitly as a function of loop parameters such as natural frequency f_n, damping factor ζ and gain factors of building blocks such as phase detector or voltage-controlled oscillator (VCO). Such equations enable the designer to tailor his/her device in order to fulfill a number of given requirements, e.g., locking within, say, 20 µs.

It is surprising that this dynamic analysis has never been performed for the Costas loop, although it has been described in many textbooks and papers. A possible reason for that could be the higher complexity of mathematics. When I tried first to develop such design equations for the Costas loop, I got aware that the Costas loop presents more nonlinearities than the PLL, which complicates the mathematical treatment considerably. Only after introducing a number of simplifications and linearizations, I was finally able to get explicit mathematical expressions for lock range, lock time, pull-in range, and pull-in time for the Costas loop. The mathematical treatment is even more aggravated because different analyses must be performed for the different types of Costas loop. The corresponding design equations will be presented in this textbook.

Another aspect of the Costas loop overlooked by almost all authors is the design of "modified" Costas loops, i.e., of Costas loops that operate with so-called pre-envelope signal, also referred to as "analytical" signal.

Operating with the pre-envelope has a dramatic impact on the performance of the Costas loop. First, it is easily shown that the lowpass filters used in conventional Costas loops are no longer required. It can be demonstrated that this greatly improves the dynamic performance of the loop, i.e., the pull-in range of such modified loops becomes much larger. When the loop filter is implemented by a PI filter (proportional + integral filter), the pull-in range becomes even infinite. Of course, this is only of "academic" interest; however in a real circuit, the loop can lock onto every frequency that can be generated by the local oscillator.

Another promising technology that has been widely discarded by most authors is the use of "phasor rotators" in Costas loops. In such systems, the local oscillator is not realized as an oscillator whose frequency can be controlled by a control signal, but as a simple oscillator that generates a constant frequency. In order to get locked, the two output signals of the Costas loop—it will be shown that there are two such signals in each Costas loop—are considered to form a "phasor," a complex quantity. Acquisition is obtained by a rotation of that phasor. Such a design offers some advantages: the complexity of the high-frequency portion is reduced, and the rotating circuits are easily implemented from some logic circuits.

The book is organized as follows:

Chapter 2 gives a short introduction to the Costas loop. It concentrates on the differences between phase-locked loop and Costas and shows by some simple examples where the PLL can be used and where the PLL fails to do the required job and should be replaced by the Costas loop.

Chapters 3 and 4 discuss the conventional Costas loops for BPSK and QPSK, where "conventional" means a loop that operates with real input signals and not with the pre-envelope signal. After theoretical investigation, design procedures are presented, including case studies for the design of analog and digital circuits.

Finally, Simulink models are shown (all MATLAB files on attached CD), which enable the designer to verify the design.

In Chaps. 5 and 6 modified Costas loops for BPSK and QPSK are discussed, including design procedure and simulation. These systems operate with the pre-envelope signal.

Chapter 7 presents theory and design of a Costas loop for m-ary PSK demodulation, with design procedure and simulation.

Chapter 8 presents Costas loop for BPSK using phasor rotation circuit with design procedure, case study for designing a digital Costas loop, and simulation.

Chapter 9 presents Costas loop for QPSK using phasor rotation circuit with design procedure, case study for designing a digital Costas loop, and simulation.

Chapter 10 presents Costas loop for demodulation of quadrature amplitude modulation (QAM) signals with theory, design procedure, and simulation.

Oberwil, Switzerland   Roland Best

Contents

1 Simulink Models   1
    1.1   Model Overview   1
    1.2   A Note on MATLAB/Simulink File Types   2
    1.3   Downloading and Installing the Simulink Models   2

2 Introduction: From Phase-Locked Loop to Costas Loop   5
    References 12

3 Conventional Costas Loop for BPSK, Dynamic Analysis,
    Design Procedure, and Simulation 13
    3.1   The Linear Model of the Costas Loop 14
    3.2   Lock Range $\Delta\omega_L$ and Lock Time T_L 18
    3.3   Nonlinear Model of the Costas Loop 19
    3.4   Pull-in Range $\Delta\omega_P$ and Pull-in Time T_P 26
    3.5   Design Procedures for Conventional Costas Loop
            for BPSK 29
            3.5.1   Case Study 1: Designing an Analog Costas
                     Loop for BPSK 29
            3.5.2   Case Study 2: Designing a Digital Costas
                     Loop for BPSK 31
    3.6   Simulating the Costas Loop for BPSK 32
    References 34

4 Conventional Costas Loop for QPSK 35
    4.1   Linear Model and Frequency Response 35
    4.2   Lock Range $\Delta\omega_L$ and Lock Time T_L 37
    4.3   Nonlinear Model in the Unlocked State 39
    4.4   Pull-in Range $\Delta\omega_P$ and Pull-in Time T_P 40
    4.5   Design Procedure for Costas Loop for QPSK 43
    4.6   Simulating Costas Loops for QPSK 46
    References 47

5   Modified Costas Loop for BPSK 49
   5.1   Linear Model and Frequency Response 50
   5.2   Lock Range $\Delta\omega_L$ and Lock Time T_L 54
   5.3   Nonlinear Model for the Unlocked State 55
   5.4   Pull-in Range and Pull-in Time of the Modified Costas
       Loop for BPSK 57
   5.5   Design Procedure for Modified Costas Loop for BPSK 57
   5.6   Simulating Modified Costas Loops for BPSK 59
   References 63

6   Modified Costas Loop for QPSK 65
   6.1   Operating Principle 65
   6.2   The Transfer Function of the Modified Costas Loop
       for QPSK 67
   6.3   Lock Range $\Delta\omega_L$ and Lock Time T_L 69
   6.4   NonLinear Model for the Unlocked State 70
   6.5   Pull-in Range and Pull-in Time of the Modified Costas
       Loop for QPSK 72
   6.6   Design Procedure for Modified Costas Loop for QPSK 73
   6.7   Simulating the Digital Costas Loop for BPSK 75
   References 75

7   Costas Loop for m-ary Phase Shift Keying (mPSK) 77
   7.1   Operating Principle 77
   7.2   Transfer Function of the modified Costas Loop for mPSK 78
   7.3   Lock Range $\Delta\omega_L$ and Lock Time T_L 80
   7.4   Nonlinear Model for the Unlocked State 82
   7.5   Pull-in Range and Pull-in Time of the Modified Costas
       Loop for QPSK 84
   7.6   Design Procedure for Costas Loop for mPSK 84
   7.7   Simulink Model for Costas Loop for mPSK 86
   References 91

8   Costas Loop for BPSK Using Phasor Rotator Circuit 93
   8.1   Operating Principle 93
   8.2   Design Procedure for Costas Loop Using Phasor Rotator 99
   8.3   Simulating the Costas Loop for BPSK Using Phasor Rotator ... 100
   8.4   Modified Costas Loop for BPSK Using Phasor Rotator 103
   8.5   Simulating the Modified Costas Loop for BPSK Using
       Phasor Rotator 104
   References 106

9   Costas Loop for QPSK Using Phasor Rotator Circuit 107
   9.1   Operating Principle 107
   9.2   Design Procedure for Costas Loop for QPSK Using
       Phasor Rotator 110

9.3   Simulating the Costas Loop for QPSK Using Phasor
        Rotator.. 111
9.4   Modified Costas Loop for QPSK Using Phasor Rotator 113
9.5   Simulating the Modified Costas Loop for QPSK Using
        Phasor Rotator 114
Reference... 116

10   **Costas Loop for Quadrature Amplitude Modulation** 117
10.1   QAM Signal Generation 117
10.2   Structure of a Costas Loop for QAM..................... 124
        10.2.1 Nyquist Filtering of Input Signal S(T)
                (RRCF1 and RRCF2)........................... 124
        10.2.2 Automatic Gain Control (AGC) 125
        10.2.3 Phase Detector............................... 127
        10.2.4 Estimator (Estim) 128
        10.2.5 Clock Recovery 128
        10.2.6 LF (Loop Filter).............................. 132
        10.2.7 Voltage-Controlled Oscillator (VCO)............. 132
10.3   Design Procedure for Costas Loop for QAM 133
        10.3.1 Blocks RRCF1 and RRCF2 (Root Raised
                Cosine Filters) 133
        10.3.2 Clock Recovery 134
        10.3.3 Frequency Control Loop (Blocks RRCF,
                Phase Detector, LF, VCO) 138
        10.3.4 Phase Detector............................... 142
        10.3.5 Preamble and Acquisition Process 143
        10.3.6 Automatic Gain Control (AGC) 144
10.4   Simulating the Costas Loop for QAM 145
        10.4.1 Simulations with Model QAM16_Nyq_mod1C 150
References 151

Index 153

Chapter 1
Simulink Models

1.1 Model Overview

In Chaps. 3–10 Simulink models are cited that can be used to simulate the performance of all types of Costas loops discussed in these chapters. The files for these models are not distributed with the book, but can be downloaded from https://doi.org/10.1007/978-3-319-72008-1_1.

The following Simulink models are available:

Chapter no.	Model filename	Type of costas loop simulated
3	BPSK_Real.mdl	Conventional costas loop for BPSK
4	QPSK_Real.mdl	Conventional costas loop for QPSK
5	BPSK_Comp.mdl	Modified costas loop for BPSK
5	BPSK_Com_PreAmb. mdl	Modified costas loop for BPSKusing preamble
6	QPSK_Comp.mdl	Modified costas loop for QPSK
7	mPSK_Comp.mdl	Modified costas loop for m-ary PSK
8	BPSK5.mdl	Costas Loop for BPSK using phasor rotator
8	BPSK6.mdl	Modified costas loop for BPSK using phasor rotator
9	QPSK5.mdl	Costas loop for QPSK using phasor rotator
9	QPSK6.mdl	Modified costas loop for QPSK using phasor rotator
10	QAM16_Nyq_mod1C. mdl	Costas loop for QAM

Electronic supplementary material The online version of this chapter (doi:10.1007/978-3-319-72008-1_1) contains supplementary material, which is available to authorized users.

1.2   A Note on MATLAB/Simulink File Types

As shown in the table above, every Simulink model is realized by a.mdl file (Simulink model file). Some of the models are also using a number of callback functions (.m files). These functions are called at different instants during a simulation. A preload function is called before the .mdl file is loaded. Some models also use init functions. These functions are called whenever the operator starts a simulation. Some models also include close functions that are called when the model is closed.

Some of the models store the parameters specified by the operator in a parameter file (.mat file). When the operator loads that model another time, the last entered parameters are restored. When operating these models, the operator must be authorized to change the content of such files. In some never versions, only administrators are allowed to modify such files. When such versions are used, the operator should choose the option "Run as administrator" when starting MATLAB.

All files of the mentioned models have been archived in one single zip file named Costas_Loops_Simulink_Files.zip. To get all Simulink models, it is only required to download that zip file and dearchive it. The next paragraph explains how to proceed.

1.3   Downloading and Installing the Simulink Models

The procedure for downloading the Simulink model is most easily explained when a number of assumptions are made first. It is assumed that

- the drive name of the hard disk in your computer is C.
- MATLAB has been installed in a folder C:\MATLAB
  (of course, any other folder name can be used).
- there exists a subfolder C:\MATLAB\WORK, where you store applications (mdl files, m files, and the like) you created yourself or downloaded from elsewhere.
  (of course, any other folder name can be used).
- the files you downloaded are stored in a folder C:\downloads
  (of course, any other folder name can be used).

Installation starts with the download of file Costas_Loops_Simulink_Files.zip. This file can be downloaded from the website https://doi.org/10.1007/978-3-319-72008-1_1.

- store the zip file in folder C:\downloads.
- open Windows Explorer and double-click the file Costas_Loops_ Simulink_Files.zip
- this opens the Winzip program. A window is displayed on the desktop. In the title bar, "Winzip Costas_Loops_Simulink_Files.zip" is displayed. In the pane

below the menu bar, all files contained in the zip archive are listed (mdl files, m files, etc).

(if you do not have Winzip installed on your computer, you can download it without any cost from the Web).

- create a subfolder C:\MATLAB\WORK\Costas_Models. This will be used to store the Simulink models

  (of course, you can specify any other subfolder name)
- select all dearchived files in the Winzip window and drag them to subfolder C: \MATLAB\WORK\Costas_Models. You are now ready to run the simulations.
- start MATLAB. Make subfolder C:\MATLAB\WORK\Costas_Models your current folder. Start a model by double-clicking the corresponding mdl file.
- for each model, a detailed description is available. To display the description, load a model and click the File menu. This brings up a pull-down menu. Click menu item "Model Properties." The model description contains a number of hints for how to run simulations, how to change model parameters, and much more.

Chapter 2
Introduction: From Phase-Locked Loop to Costas Loop

The Costas loop can be considered an extended version of the phase-locked loop (PLL). The PLL has been invented in 1932 by French engineer Henri de Belleszice [1]. In his first application, de Belleszice used the PLL as a synchronous demodulator for double sideband amplitude modulated signals with carrier. The block diagram of a PLL is shown in Fig. 2.1. It is built from three blocks, a phase detector (PD), a loop filter (LF), and a voltage-controlled oscillator (VCO). In the first PLL applications, an analog multiplier was used for the phase detector [2]. Assuming for the first moment that both signals U_1 and U_2 are sinusoidal, we can write

$$u_1(t) = U_{10} \sin(\omega_1 t + \theta_1)$$
$$u_2(t) = U_{20} \cos(\omega_1 t + \theta_2)$$

where U_{10}, U_{20} are the amplitudes of U_1 and U_2, respectively, ω_1 is the radian frequency of the input signal U_1, and θ_1 and θ_2 are the zero phases of U_1 and U_2, respectively. Assume further that the system is already locked, i.e., both signals have the same frequency, but can have different phases. In this case, the signals u_1 and U_2 differ by 90° in the locked state; hence, it is reasonable to define U_1 as a sine wave and U_2 as a cosine wave. It can be shown that the output signal of the phase detector is proportional to $\sin(\theta_1 - \theta_2) = \sin(\theta_e)$, where θ_e is called phase error. But the output signal of that type of phase detector also contains a high-frequency term, i.e., a sine term having radian frequency 2 ω_1. This term is removed by the loop filter, which is mostly realized either as a lag-lead filter or as a PI filter (proportional

Electronic supplementary material The online version of this chapter (doi:10.1007/978-3-319-72008-1_2) contains supplementary material, which is available to authorized users.

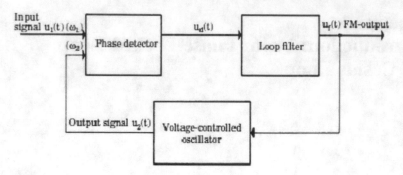

Fig. 2.1 Block diagram of a phase-locked loop

and integral filter). More about loop filters later in this text. The output signal u_f of the loop filter is applied to the input of the VCO. When there exists a phase error, the frequency of the VCO is adjusted such that finally the phase error becomes either 0 or is at least very small.

Next we consider a PLL circuit used for synchronous demodulation of AM signals.

Figure 2.2 shows the relevant signals. The upper trace is the modulating signal. It is scaled such that it is within the range from −1 to 1. The middle trace is the carrier signal c(t). The modulated signal U_1 is given by

$$u_1 = c(t)(1 + m \, u_m(t))$$

with m = modulation index. m must be chosen m < 1, a commonly used value is m = 0.3. In this case, the modulated signal u_1 (lower trace) is always in-phase with the carrier c(t). If m were chosen larger than 1, U_1 could be in antiphase with the carrier when the modulating signal has large negative values. When the modulated signal u_1 is now applied to the input of a PLL, the output signal U_2 of the VCO would correctly lock onto that signal, i.e., there would always be a phase difference of 90° between u_1 and u_2. Figure 2.3 shows the block diagram of a PLL designed for synchronous demodulation of the amplitude modulated signal.

Four blocks have been added to the basic circuit of Fig. 2.1, a 90° phase shifter, a multiplier (MUL), a lowpass filter (LPF), and a highpass filter (HPF). As we have seen, there is a phase difference of 90° between U_1 and U_2 when the PLL has locked. When U_2 is shifted by 90°, the shifted signal $U_{2,shift}$ is exactly in-phase with the modulated carrier U_1. With the definitions

$$u_1 = U_{10} \sin(\omega_1 + \theta_1)(1 + m \, u_m)$$
$$u_{2,shift} = U_{20} \sin(\omega_1 + \theta_1)$$

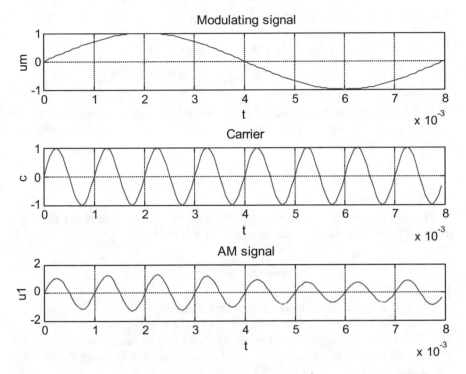

Fig. 2.2 Double sideband AM with carrier. Upper trace: modulating signal U_m, middle trace: carrier c, lower trace: modulated signal U_1

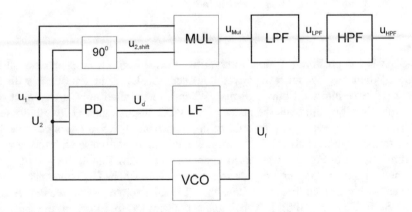

Fig. 2.3 PLL used for synchronous demodulation of AM signal

the output signal of the multiplier U_{Mul} becomes

$$u_{Mul} = U_{10}U_{20}(1 + m\, u_m)\sin^2(\omega_1 t + \theta_1)$$
$$= U_{10}U_{20}(1 + m\, u_m)\left[\frac{1}{2} - \frac{1}{2}\cos(2\omega_1 t + 2\theta_1)\right]$$

We recognize that U_{Mul} contains a high-frequency term centered at twice the center radian frequency ω_1. This term is removed from the lowpass filter; hence, its output signal is given by

$$u_{LPF} = \frac{U_{10}U_{20}}{2}(1 + m\, u_m)$$

This signal contains a dc term $U_{10}U_{20}/2$. This can be removed if required by a highpass filter. The output of the highpass filter is then

$$u_{HPF} = \frac{U_{10}U_{20}m}{2}u_m$$

which is identical with the modulating signal scaled by a factor $U_{10}U_{20}\, m/2$.

Whereas the PLL can be successfully used for the synchronous demodulation of double sideband AM signals with carrier, it fails when it cames to demodulated AM signals with suppressed carriers. The waveforms of such an AM signal are shown in Fig. 2.4.

First trace: modulating signal U_m. Second trace: carrier $c(t)$. Third trace: modulated signal U_1. Fourth trace: reconstructed carrier $U_{2,shift}$.

The modulated signal U_1 is given here by

$$u_1 = u_m \sin(\omega_1 t)$$

When U_m is positive (cf. time interval from $0...4$ ms), U_1 is in-phase with the carrier. When U_m becomes negative, however, the U_1 is in antiphase with the carrier (cf. time interval from $4...8$ ms). When U_1 is applied now to the input of a PLL, this circuit would track the phase of the VCO output signal U_2 to the phase of U_1. The shifted signal $U_{2,shifted}$ would be in-phase with U_1 when U_m is positive, but after a transient in the interval $4...5$ ms, it would lock in antiphase with the carrier c (t). If the circuit in Fig. 2.3 were used to demolate the AM signal, the output signal U_{Mul} would have wrong polarity during the intervals where U_m is negative.

There is another application where the PLL fails for the same reason: Binary Phase Shift Keying (BPSK). The signals are similar to those in the previous example, as shown in Fig. 2.5.

The same happens as in the previous example. Because the polarity of the BPSK signal is reversed when the binary signal becomes negative, the reconstructed carrier $U_{2,shift}$ is in antiphase with the carrier $c(t)$, when the modulating signal U_m is negative.

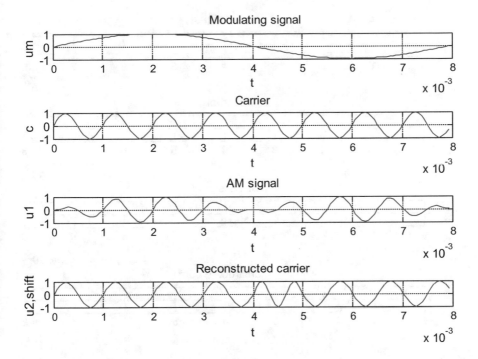

Fig. 2.4 Waveforms of a double sideband AM signal with suppressed carrier

Here the Costas loop comes into play [3]. Figure 2.6 shows the block diagram of the conventional Costas loop for BPSK. Compared with the PLL, this novel circuit consists of two branches, the I branch and the Q branch, whereas the PLL has only one. The input signal is given by $U_1 = m(t) \sin(\omega_1 t + \theta_1)$, where m is the data signal and can two values, +c or –c, where c is a constant. In many cases, c = 1 is chosen. When the currently transmitted bit is a logical one, m = + c, and when the currently transmitted bit is a logical zero, m = –c. The voltage-controlled oscillator in this circuit has two outputs that differ by 90° in-phase, i.e., a sine output and a cosine output. The input signal is multiplied by the sine wave in the I branch, and it is multiplied by the cosine wave in the Q branch. Consequently, the output of the multiplier in the I branch becomes $m(t) \sin(\omega_1 t + \theta_1) \, 2 \sin(\omega_1 t + \theta_2)$. This signal contains a high-frequency term whose frequency is centered around $2\,\omega_1$. This term is removed by lowpass filter LPF1, and the output signal of this filter becomes $m(t) \cos(\theta_1 - \theta_2) = m(t) \cos\theta_e$, where θ_e is the phase error. In analogy, the output signal of lowpass filter LPF2 becomes $m(t) \sin\theta_e$. Now the output signals of both lowpass filters are multiplied; hence, the output signal of multiplier MUL is

$$u_d = \frac{m^2}{2} \sin(2\,\theta_e)$$

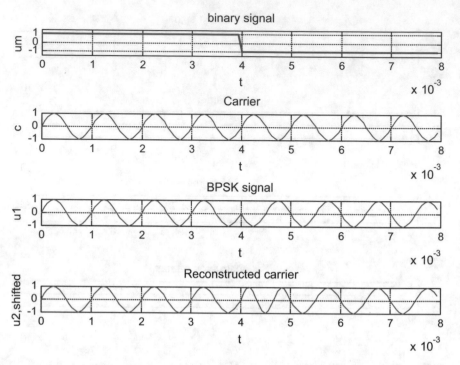

Fig. 2.5 Waveforms for BPSK. First trace: binary signal u_m. Second trace: carrier $c(t)$. Third trace: BPSK signal u_1. Fourth trace: reconstructed carrier $U_{2,shift}$, shifted by $90°$

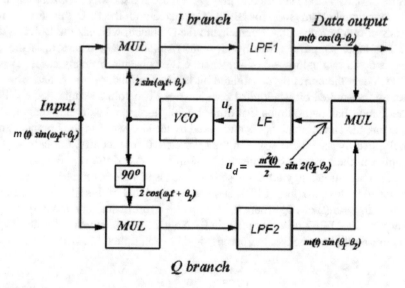

Fig. 2.6 Block diagram of Costas loop for BPSK

When the phase error is small, this can be written

$$u_d = m^2 \theta_e$$

i.e., the multiplier represents a phase detector having detector gain

$$K_d = m^2$$

We recognize that the output signal u_d of the phase detector does not depend on the polarity of signal m, due to the factor m^2. Thus, the phase detector output signal does not change polarity when m changes from positive to negative values and vice versa. The output signal u_d of the multiplier (phase detector) is applied to the input of the loop filter LF. This filter is always realized as a lowpass filter. Two kinds of loop filters are in use, the lag-lead filter and the PI filter (proportional and integral filter) [2, 4]. When there is a phase error θ_e, the output signal of the loop filter controls the frequency of the VCO such that the phase error is reduced to zero or to a very small value. When the phase error has been reduced to zero or near zero, the output signal of lowpass filter LPF1 is identical with the data signal m(t).

We have seen that the Costas loop can adjust the frequency and phase of the VCO such loop locks with a phase difference $\theta_1 - \theta_2$ near zero. It should be noted that the Costas loop can also lock with a phase difference $\theta_1 - \theta_2$ of π. Assume for the moment that $\theta_1 - \theta_2$ has not yet attained the value π, but is near π. We then can set

$$\theta_1 - \theta_2 = \pi + \theta_e$$

Under this condition, the output signal of LPF1 becomes m(t) cos($\theta_1 - \theta_2$) = −m (t) cos θ_e, and the output signal of LPF2 becomes −m(t) sin θ_e. The output signal of the multiplier then becomes again

$$u_d = m^2 \theta_e$$

Here again, when there exists a phase error, the frequency of the VCO will be adjusted such that the loop stably locks with a phase difference $\theta_1 - \theta_2 = \pi$. We can conclude therefore that the Costas loop can lock at two different points of equilibrium, i.e., with $\theta_1 - \theta_2 = 0$ or with $\theta_1 - \theta_2 = \pi$. When the loop locks with a phase difference of π, the output signal of lowpass filter LPF1 becomes −m, i.e., its polarity gets inverted.

This is not necessarily a problem, because in many cases differential encoding is used with BPSK [5]. With standard—i.e., not differential—encoding the value of the currently transmitted bit depends only on the polarity of signal m(t). The Costas loop can decide that a transmitted bit is a logical 1 when m is positive or a logical 0 when m is negative. When differential encoding is applied, the value of the currently transmitted bit depends from two values, i.e., from the polarity of the current bit and the polarity of the previously transmitted bit. We define, for example, that the currently transmitted bit is a logical 1 when current and previous bit have opposite polarity, and that the value of that bit is a logical 0 when these two samples have the same polarity. Under this condition, the Costas loop can lock onto any of the two equilibrium states.

Non-differential encoding can be used when the Costas loop can be brought to lock a priori with the "correct" phase difference of 0. Assume that a transmitter starts to send a series of binary data, e.g., a series of 256 bits. To obtain correct locking, a preamble is preceding the data block, e.g., a series of 16 logical 1's. The Costas loop must now be equipped with an initialization circuit that becomes active at start of data transmission. Because the Costas loop "knows" the correct value of the first received bits, the initialization circuit can control the VCO such that false locking (i.e., locking with a phase difference of π is prevented). In Sect. 5.6, an example of a Costas loop using such a preamble is presented.

References

1. De Belleszice H., La Reception Synchrone, L'onde electrique, **11,** 225–240 (1932)
2. E. Roland, *Best, phase-locked loops, design, simulation, and applications*, 6th edn. (McGraw-Hill, New York, 2007)
3. J.P. Costas, Synchronous communications. Proc. IRE. 1713–1718 (1956)
4. U. Rohde, J. Whitacker, *Communications receivers* (Software Radios, and Design, McGraw-Hill, DSP, 2001)
5. B. Sklar, *Digital communications, fundamentals and applications* (Prentice Hall, USA, 1988)

Chapter 3
Conventional Costas Loop for BPSK, Dynamic Analysis, Design Procedure, and Simulation

As we have seen in Chap. 2, the Costas loop can be considered a variant of the phase-locked loop (PLL). Both systems are synchronizing devices. When a data transmission starts, the loop has not yet acquired lock, i.e., there is a difference between the carrier radian frequency ω_1 of the input signal and the radian frequency ω_2 of the local oscillator (VCO, cf. Fig. 3.1). A process called acquisition process will then start. In case of the PLL, two different acquisition processes have been defined [2, 4, 5]. When the difference between carrier frequency and initial frequency of the local oscillator is within a range called lock range $\Delta\omega_L$ the acquisition process is relatively fast, and the loop will lock within at most one beat note between the frequencies ω_1 and ω_2. The time required to get locked is called lock time T_L. When the difference between carrier frequency and initial frequency of the local oscillator is outside the lock range but inside another range called pull-in range $\Delta\omega_P$, a slower acquisition process will take place. There will be a number of beat notes between the two frequencies, and the time required to get locked, the pull-in time T_P, is much longer than T_L. For the PLL, equations have been developed that enable the designer to calculate all these parameters, $\Delta\omega_L$, T_L, $\Delta\omega_P$, and T_P from the properties of the PLL such as natural frequency, damping factor, phase detector gain K_d, VCO gain K_0. Astonishingly, this analysis has never been made for the Costas loop. This will therefore be done in the following sections. As it will show up the parameters for the lock process $\Delta\omega_L$, T_L can be derived from the parameters of the phase transfer function H(s) of the Costas loop. To derive equations for the parameters of the pull-in process, however, we will need more complex mathematical models, i.e., models for the unlocked loop.

Electronic supplementary material The online version of this chapter (doi:10.1007/978-3-319-72008-1_3) contains supplementary material, which is available to authorized users.

3.1   The Linear Model of the Costas Loop

First we develop a linear model for this type of Costas loop that is valid in the locked state of the loop. The block diagram of the Costas loop is shown in Fig. 3.1a.

The input signal is given by

$$u_1(t) = m_1(t) \sin(\omega_1 t + \theta_1) \tag{3.1}$$

where $m_1(t)$ is the data signal, ω_1 is the carrier frequency, and θ_1 is the zero phase. In many cases, $m_1(t)$ has either the value 1 or -1 but it also can have arbitrary values. As Gardner [1] has shown the output signal of the multiplier at the right of the block diagram represents the phase detector output signal and is given by

$$u_d(t) = \frac{m_1^2}{2} \sin 2\theta_e, \ with \ \theta_e = \theta_1 - \theta_2 \tag{3.2}$$

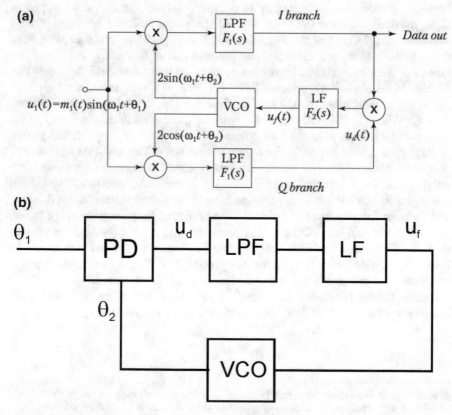

Fig. 3.1 **a** Block diagram of Costas loop for BPSK. **b** Definition of symbols for the Costas loop

In the locked state, the phase error θ_e is very small so we can write

$$u_d(t) \approx m_1^2 \theta_e = K_d \theta_e \tag{3.3}$$

with K_d called detector gain. For the locked state, a simplified block diagram can be derived therefore, which is shown in Fig. 3.1b.

In the locked state, both reference and VCO frequencies are approximately the same; hence, the input of the lowpass filter is a very low frequency signal. Therefore, the lowpass filter can be ignored when setting up the linear model of the Costas loop. The linear model is made up of three blocks: the phase detector PD, the loop filter LF, and the VCO (voltage-controlled oscillator). In digital Costas loops the VCO is replaced by a digital-controlled oscillator (DCO). This will be discussed in Sect. 3.5.2. For these building blocks, the transfer functions are now defined as follows:

Phase detector PD

$$H_{PD}(s) = \frac{U_d(s)}{\Theta_e(s)} = K_d \tag{3.4}$$

Note that the uppercase symbols are Laplace transforms of the corresponding lower case signals.

Lowpass filter LPF

As described in Chap. 2, the lowpass filter is used to suppress the high frequency component centered around twice the carrier frequency ($2\,\omega_1$). This is usually a first-order lowpass filter with corner radian frequency ω_3; hence, its transfer function is given by

$$H_{LPF}(s) = \frac{1}{1 + s/\omega_3}$$

Loop filter LF

For the loop filter, we choose a PI (proportional + integral) filter whose transfer function

$$H_{LF}(s) = \frac{U_f(s)}{U_d(s)} = \frac{1 + s\tau_2}{s\tau_1} \tag{3.5}$$

This filter type is the preferred one because it offers superior performance compared with lag or lag–lead filters.

VCO

The transfer function of the VCO is given by

$$H_{VCO}(s) = \frac{\Theta_2(s)}{U_f(s)} = \frac{K_0}{s} \qquad (3.6)$$

where K_0 is called VCO gain.

We now can derive the open loop transfer function of the Costas loop which is defined by the ratio $\Theta_2(s)/\Theta_1(s)$

$$G_{OL}(s) = K_d \frac{K_0}{s} \frac{1 + s\tau_2}{s\tau_1} \cdot \frac{1}{1 + s/\omega_3} \qquad (3.7)$$

Figure 3.2 shows a Bode plot of the magnitude of G_{OL}. The plot is characterized by the corner frequency ω_C which is defined by $\omega_C = 1/\tau_2$, and gain parameters K_d and K_0. At lower frequencies, the magnitude rolls off with a slope of −40 dB/decade. At frequency ω_C, the zero of the loop filter causes the magnitude to change its slope to −20 dB/decade. To get a stable system, the magnitude curve should cut the 0 dB line with a slope that is markedly less than −40 dB/decade. Setting the parameters such that the gain is just 0 dB at frequency ω_C provides a phase margin of 45° which assures stability [2]. At radian frequency ω_3, the influence of the lowpass filter becomes visible. Above ω_3, the slope of the open loop transfer function switches to −40 dB/decade. The placement of the pole at s = ω_3 is critical. When it was placed at lower frequencies, e.g., below corner frequency ω_C, the slope would approach −60 dB/decade which would make the system unstable. It is therefore mandatory to place the pole at a frequency where the open loop gain is markedly below 0 dB. When doing so, the pole at s = ω_3 has almost no effect onto the dynamic performance of the loop and can therefore be discarded for the following analysis. We will see later, however, that the pole at s = ω_3 will come into play when nonlinear mathematical models for the Costas loop are developed.

From the open loop transfer function, we now can calculate the closed loop transfer function defined by

$$G_{CL}(s) = \frac{\Theta_2(s)}{\Theta_1(s)} \qquad (3.8)$$

After some mathematical manipulations, we get

$$G_{CL}(s) = \frac{K_0 K_d \frac{1 + s\tau_2}{s\tau_1}}{s^2 + s\frac{K_0 K_d \tau_2}{\tau_1} + \frac{K_0 K_d}{\tau_1}} \qquad (3.9)$$

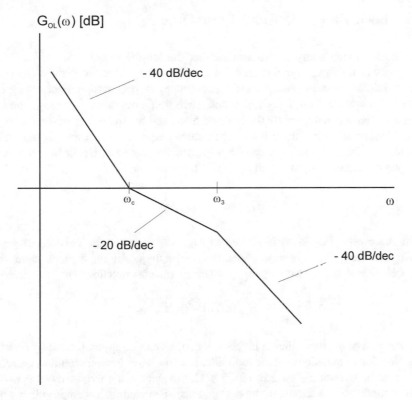

Fig. 3.2 Bode plot of magnitude of open loop gain $G_{OL}(\omega)$

It is customary to represent this transfer function in normalized form, i.e.,

$$G_{CL}(s) = \frac{2s\zeta\omega_n + \omega_n^2}{s^2 + 2s\zeta\omega_n + \omega_n^2} \tag{3.10}$$

with the substitutions

$$\omega_n = \sqrt{\frac{K_0 K_d}{\tau_1}} \quad , \quad \zeta = \frac{\omega_n \tau_2}{2} \tag{3.11}$$

where ω_n is called *natural frequency* and ζ is called *damping factor*. The linear model enables us to derive simple expressions for lock range $\Delta\omega_L$ and lock time T_L.

3.2   Lock Range $\Delta\omega_L$ and Lock Time T_L

For the following analysis, we assume that the loop is initially out of lock. The frequency of the input signal (Fig. 3.1a) is ω_1, and the frequency of the VCO is ω_2. The multiplier in the I branch therefore generates an output signal consisting of a sum frequency term $\omega_1 + \omega_2$ and a difference frequency term $\omega_1 - \omega_2$. The sum frequency term is removed by the lowpass filter, and the frequency of the difference term is assumed to be much below the corner frequency ω_3 of the lowpass filter; hence, the action of this filter can be neglected for this case. Under this condition, the phase detector output signal $u_d(t)$ will have the form [cf. Eqs. (3.2) and (3.3)]

$$u_d(t) = \frac{K_d}{2} \sin\left(\Delta\omega\ t\right)$$

with $\Delta\omega = \omega_1 - \omega_2$. $u_d(t)$ is plotted in Fig. 3.3, upper trace. This signal passes through the loop filter. In most cases, the corner frequency ω_2 is much lower than the lock range; hence, we can approximate its transfer function by

$$H_{LF}(\omega) \approx \frac{\tau_2}{\tau_1} = K_H$$

i.e., the gain of the loop filter at higher frequencies can be approximated by constant K_H. Now the output signal of the loop filter is a sine wave having amplitude $K_d K_H/2$ as shown by the middle trace in Fig. 3.3. Consequently, the frequency of the VCO will be modulated as shown in the bottom trace. The modulation amplitude is given by $K_d K_0 K_H/2$. In this figure, the reference frequency and the initial frequency ω_{20} of the VCO are plotted as horizontal lines. When ω_1 and ω_{20} are such that the top of the sine wave just touches the ω_1 line, the loop acquires lock suddenly, i.e., the lock range $\Delta\omega_L$ is nothing more than the modulation amplitude $K_d K_0 K_H/2$. Making use of the substitutions (3.11), we finally get

$$\Delta\omega_L = \zeta\,\omega_n \qquad\qquad (3.12)$$

Now the lock process is a damped oscillation whose frequency is the natural frequency. Because the loop is assumed to lock within at most one cycle of that frequency, the lock time can be approximate by the period of the natural frequency, i.e., we have

$$T_L \approx \frac{2\pi}{\omega_n} \qquad\qquad (3.13)$$

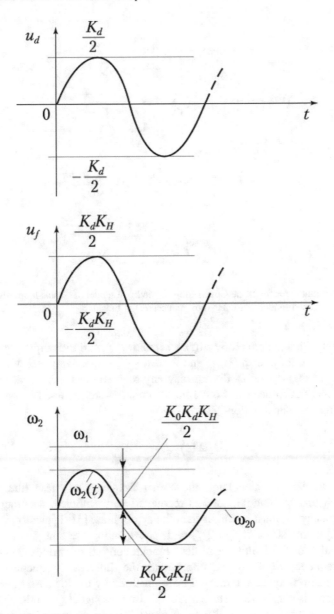

Fig. 3.3 Lock range of Costas loop

3.3 Nonlinear Model of the Costas Loop

We need a nonlinear model to derive expressions for pull-in range $\Delta\omega_P$ and pull-in time T_P. Let us start with the block diagram in Fig. 3.4a.

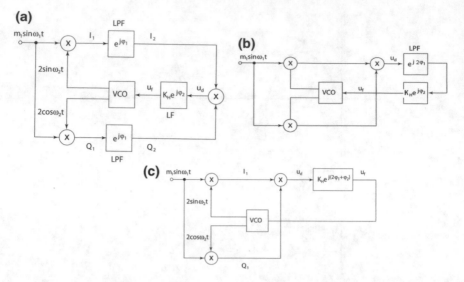

Fig. 3.4 **a** Nonlinear model of the Costas loop. **b** Modified model of Costas loop, reversed order of blocks. **c** modified model of Costas loop, concatenated blocks

The input signal is defined by $u_1(t) = m_1 \sin(\omega_1 t)$, and the output signals of the VCO are defined to be $2 \sin(\omega_2 t)$ and $2 \cos(\omega_2 t)$, respectively. As will be shown later, initial phases θ_1 and θ_2 do not play any role here and can be discarded. The lowpass filters (LPF) used in both I an Q branches are assumed to be first-order filters having transfer function

$$H_{LPF}(s) = \frac{1}{1 + s/\omega_3} \qquad (3.14)$$

As will be demonstrated later, the corner frequency of these filters must be chosen such that the data signal I is recovered with sufficient accuracy, i.e., the corner frequency ω_3 must be larger than the symbol rate [3]. Typically, it is chosen twice the symbol rate, i.e., $f_3 = 2 f_S$ with f_S = symbol rate and $f_3 = \omega_3/2 \pi$. The output signal I_1 of the multiplier in the I branch consists of two terms: one having the sum frequency $\omega_1 + \omega_2$ and one having the difference frequency $\omega_1 - \omega_2$. Because the sum frequency term will be suppressed by the lowpass filter, only the difference term is considered. The same holds true for signal Q_1 in the Q branch. It will show up that the range of difference frequencies is markedly below the corner frequency ω_3 of the lowpass filter. Hence, the filter gain will be nearly 1 for the frequencies of interest. As will also be shown later the phase, however, at frequency $\Delta\omega = \omega_1 - \omega_2$ cannot be neglected. The lowpass filter is therefore represented as a delay block whose transfer function has the value $exp(j\ \varphi_1)$, where φ_1 is the phase at frequency $\Delta\omega$. The delayed signals I_2 and Q_2 are now multiplied by the product block at the right in the block diagram. Consequently, the output signal $u_d(t)$ of this block will have a frequency of 2 $\Delta\omega$.

This signal is now applied to the input of the loop filter LF. Its transfer function has been defined in Eq. (3.5). The corner frequency of this filter is $\Delta\omega_C = 1/\tau_2$. From Eqs. (3.11) and (3.12), it can be seen that ω_C is in the order of the lock range $\Delta\omega_L$. Because we are considering here difference frequencies $\Delta\omega$ that are beyond the lock range, $\Delta\omega$ is markedly larger than the corner frequency $\Delta\omega$, and the gain K_H of the loop filter can be approximated by

$$K_H \approx \frac{\tau_2}{\tau_1} \tag{3.15}$$

Because the phase of the loop filter cannot be neglected, it is represented as a delay block characterized by

$$H_{LF}(2\,\Delta\omega) = K_H \exp(j\varphi_2) \tag{3.16}$$

where φ_2 is the phase of the loop filter at frequency 2 $\Delta\omega$.

The analysis of dynamic behavior becomes easier when the order of some blocks in Fig. 3.4a is reversed, i.e., when we put the multiplying block before the lowpass filter [3].

The modified block diagram is shown in Fig. 3.4b. Because the frequency of signal $u_d(t)$ in Fig. 3.4a is twice the frequency of the signals I_2 and Q_2, the phase shift created by the lowpass filter at frequency 2 $\Delta\omega$ is now twice the phase shift at frequency $\Delta\omega$. The LPF is therefore represented here by a delay block having transfer function $exp(2\,j\,\varphi_1)$.

We can simplify the block diagram even more by concatenating the lowpass filter and loop filter blocks. The resulting block delays the phase by $\varphi_{tot} = 2\,\varphi_1 + \varphi_2$. This is shown in Fig. 3.4c. The output signal $u_f(t)$ of this delay block now modulates the frequency generated by the VCO.

We have seen that all signals found in this block diagram are sine functions, i.e., all of them seem to have zero average, hence do not show any dc component. This would lead to the (erroneous) conclusion that a pull-in process would not be possible. In reality, it will be recognized that some of the signals become asymmetrical, i.e., the duration of the positive half wave is different from the duration of the negative one. This creates a nonzero dc component, and under suitable conditions acquisition can be obtained. We are therefore going to analyze the characteristics of the signals in Fig. 3.4c.

All considered signals are plotted in Fig. 3.5. For signals I_1 and Q_1, we obtain

$$I_1(t) = m_1 \cos(\Delta\omega\, t)$$
$$Q_1(t) = m_1 \sin(\Delta\omega\, t)$$

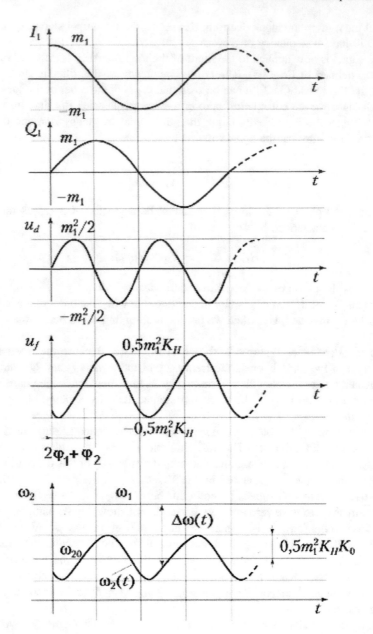

Fig. 3.5 Signals of the model in Fig. 3.4c

Note that only the difference frequency terms are considered. The sum frequency terms are discarded because they are removed by the lowpass filter. The signal $u_d(t)$ is the product of I_1 and Q_1 and is given by

$$u_d(t) = \frac{m_1^2}{2} \sin(2 \, \Delta\omega \, t)$$

Next the loop filter output signal $u_f(t)$ is plotted. Its amplitude is $K_H \, m_1^2/2$, and its phase is delayed by $\varphi_{tot} = 2 \, \varphi_1 + \varphi_2$. This signal modulates the frequency of the VCO as shown in the bottom trace of Fig. 3.5. The modulation amplitude is given by $0.5 \, m_1^2 \, K_H \, K_0$. In order to get an estimate for the nonzero dc component of $u_d(t)$, we will have to analyze the asymmetry of the signal waveforms. It will be shown that $\overline{u_d}$ [the average of $u_d(t)$] is a function of frequency difference $\Delta\omega$ and phase φ_{tot}. The analysis becomes easier when we first calculate $\overline{u_d}$ for some special values for φ_{tot}, i.e., for $\varphi_{tot} = 0$; $-\pi/2$; and $-\pi$. Let us start with $\varphi_{tot} = 0$, cf. Fig. 3.6a.

Shown are the waveforms for $u_d(t)$ and $\omega_2(t)$. The asymmetry of the signals is exaggerated in this plot. During the positive half cycle (duration T_1), the average value of VCO output frequency $\omega_2(t)$ is increased, which means that the average difference frequency $\Delta\omega(t)$ is lowered. Consequently, the duration of the positive half wave becomes larger than half of a full cycle. During the negative half cycle (duration T_2), however, the average value of VCO output frequency $\omega_2(t)$ is decreased, which means that that the average difference frequency $\Delta\omega(t)$ is increased. Consequently, the duration of the negative half wave becomes less than half of a full cycle.

Next we are going to calculate the average frequency difference in both half cycles. The average frequency difference during half cycle T_1 is denoted $\overline{\Delta\omega_{d+}}$, and the average frequency difference during half cycle T_2 is denoted $\overline{\Delta\omega_{d-}}$. We get

$$\overline{\Delta\omega_{d+}} = \Delta\omega - \frac{2}{\pi} \frac{K_0 K_d K_H}{2} \tag{3.17a}$$

$$\overline{\Delta\omega_{d-}} = \Delta\omega + \frac{2}{\pi} \frac{K_0 K_d K_H}{2} \tag{3.17b}$$

For the durations T_1 and T_2, we obtain after some manipulations

$$T_1 \approx \frac{\pi}{2 \, \Delta\omega} \left(1 + \frac{K_0 K_d K_H}{\pi \, \Delta\omega} \right) \tag{3.18a}$$

$$T_2 \approx \frac{\pi}{2 \, \Delta\omega} \left(1 - \frac{K_0 K_d K_H}{\pi \, \Delta\omega} \right) \tag{3.18b}$$

Now the average value $\overline{u_d}$ can be calculated from

$$\overline{u_d(t)} = \frac{K_0 K_d^2 K_H}{\pi^2 \, \Delta\omega} \tag{3.19}$$

The average signal $\overline{u_d}$ is seen to be inversely proportional to the frequency difference $\Delta\omega$. Because $\overline{u_d}$ is positive, the instantaneous frequency $\omega_2(t)$ is pulled in

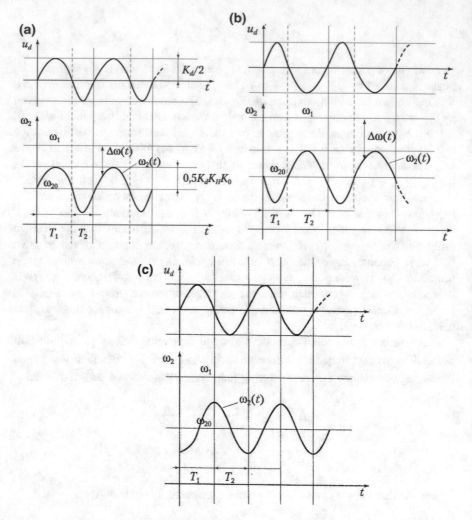

Fig. 3.6 a Signals of model in Fig. 3.4c for $\varphi_{tot} = 0$. **b** Signals of model in Fig. 3.4c for $\varphi_{tot} = -\pi$. **c** Signals of model in Fig. 3.4c for $\varphi_{tot} = -\pi/2$

positive direction, i.e., versus ω_1 which means that a pull-in process will take place. The loop will lock after a time called pull-in time T_P. This will be discussed in Sect. 3.4.

Next we are going to analyze the dependence of $\overline{u_d}$ on phase φ_{tot}. Let us consider now the case for $\varphi_{tot} = -\pi$, cf. Fig. 3.6b. We observe that in interval T_1, the instantaneous frequency $\omega_2(t)$ is pulled in negative direction; hence, the average difference frequency $\overline{\Delta\omega_{d+}}$ becomes larger. Consequently, interval T_1 becomes shorter. In interval T_2, however, the reverse is true. Here the instantaneous frequency T_1 pulled in positive direction, hence the average $\overline{\Delta\omega_{d-}}$ is reduced, and interval T_2 becomes longer. The average $\overline{u_d}$ is now equal and opposite to the value of $\overline{u_d}$ for $\varphi_{tot} = 0$. Because $\overline{u_d}$ is negative under this condition, a pull-in process cannot take place, because the frequency of the VCO is "pulled away" in the wrong direction.

Last we consider the case $\varphi_{tot} = -\pi/2$, cf. Fig. 3.6c. In the first half of interval T_1, the instantaneous frequency $\omega_2(t)$ is lowered, but in the second half it is increased. Consequently, the average difference frequency $\overline{\Delta\omega_{d+}}$ does not change its value during T_1. The same happens in interval T_2. $\overline{\Delta\omega_{d-}}$ does not change either, and $\overline{u_d}$ remains 0.

It is easy to demonstrate that $\overline{u_d}$ varies with $cos(\varphi_{tot})$; hence, we have

$$\overline{u_d(t)} = \frac{K_0 K_d^2 K_H}{\pi^2 \, \Delta\omega} cos(\varphi_{tot}), \quad \varphi_{tot} = 2\varphi_1 + \varphi_2 \qquad (3.20)$$

Equation (3.20) tells us that the pull-in range is finite. The pull-in range is the frequency difference for which phase $\varphi_{tot} = -\pi/2$. An equation for the pull-in range will be derived in Sect. 3.4. We also will have to find an equation for the pull-in time. The model shown in Fig. 3.7 will enable us to obtain a differential equation for the average frequency difference $\Delta\omega$ as a function of time.

The model is built from three blocks. The first of these is labeled "phase–frequency detector." We have seen that in the locked state, the output of the phase detector depends on the phase error θ_e. In the unlocked state, however, the average phase detector output signal $\overline{u_d}$ is a function of frequency difference as shown in Eq. (3.20); hence, it is justified to call that block "phase-frequency detector." As we will recognize the pull-in process is a slow one, i.e., its frequency spectrum contains low frequencies only that are below the corner frequency ω_C of the loop filter, cf. Eq. (3.5). The loop filter can therefore be modeled as a simple integrator with transfer function

$$H_{LF}(s) \approx \frac{1}{s \tau_1} \qquad (3.21a)$$

Fig. 3.7 Nonlinear model of Costas loop for computation of pull-in time

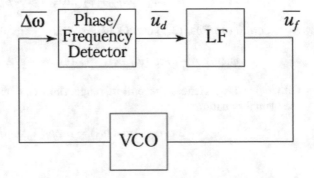

In time domain, we can therefore write

$$\overline{u_f(t)} = \frac{1}{\tau_1} \int_0^t \overline{u_d(\tau)}\, d\tau \tag{3.21b}$$

The frequency ω_2 of the VCO output signal is defined as

$$\omega_2 = \omega_{20} + K_0 \overline{u_f} \tag{3.22}$$

where ω_{20} is the free-running frequency and K_0 is the VCO gain. Now we introduce the initial frequency difference $\Delta\omega_0$ as

$$\Delta\omega_0 = \omega_1 - \omega_{20} \tag{3.23}$$

(ω_1 = carrier frequency or reference frequency) and the instantaneous frequency difference $\Delta\omega$ as

$$\Delta\omega = \omega_1 - \omega_2 \tag{3.24}$$

Substituting (3.23) and (3.24) into (3.22) finally yields

$$\Delta\omega = \Delta\omega_0 - K_0 \overline{u_f} \tag{3.25}$$

Equations (3.20), (3.21b), and (3.25) enable us to compute the three variables $\overline{u_d}$, $\overline{u_f}$ and $\Delta\omega$ as a function of time. This will be demonstrated in Sect. 3.4.

3.4 Pull-in Range $\Delta\omega_P$ and Pull-in Time T_P

The pull-in range can be computed using Eq. 3.20. Lock can only be obtained when the total phase shift φ_{tot} is not more negative than $-\pi/2$. This leads to an equation of the form

$$2\varphi_1(\Delta\omega_P) + \varphi_2(2\Delta\omega_P) = -\pi/2 \tag{3.26}$$

According to Eqs. (3.5) and (3.14), φ_1 and φ_2 are given by

$$\varphi_1(\omega) = -arctg(\omega/\omega_3), \quad \varphi_2(\omega) = -\pi/2 + arctg(\omega/\omega_C)$$

with $\omega_C = 1/\tau_2$. Hence, the pull-in range $\Delta\omega_P$ can be computed from the transcendental equation

$$2\, arctg(\Delta\omega_P/\omega_3) = arctg(2\, \Delta\omega_P/\omega_C) \tag{3.27}$$

To solve this equation for $\Delta\omega_P$, we use the addition theorem for the tangent function

$$tg(2\alpha) = \frac{2\,tg\alpha}{1 - tg^2\alpha}$$

and can replace $2\,arctg\,(\Delta\omega_P/\omega_3)$ by $arc\,tg\,\dfrac{2\frac{\Delta\omega_P}{\omega_3}}{1 - \frac{\Delta\omega_P^2}{\omega_3^2}}$. Equation (3.27) can now be

rewritten as $arc\,tg\,\dfrac{2\frac{\Delta\omega_P}{\omega_3}}{1 - \frac{\Delta\omega_P^2}{\omega_3^2}} = arc\,tg\,2\frac{\Delta\omega_P}{\omega_C}$.

When the $arc\,tg$ expressions on both sides of the equation are equal, their arguments must also be identical, which leads to

$$\frac{2\frac{\Delta\omega_P}{\omega_3}}{1 - \frac{\Delta\omega_P^2}{\omega_3^2}} = 2\,\frac{\Delta\omega_P}{\omega_C}$$

Hence, we get for the pull-in range

$$\Delta\omega_P = \sqrt{\omega_3(\omega_3 - \omega_C)} \tag{3.28}$$

It is easily seen from Eq. (3.28) that the pull-in range is positive and real only when the corner frequency ω_C of the loop filter is chosen smaller than the corner frequency ω_3 of the lowpass filters. When ω_C is chosen larger than ω_C, the loop becomes unstable. When ω_C is markedly lower than ω_3, the pull-in range approaches ω_3. When a large pull-in range is desired, a large corner frequency ω_3 of the lowpass filters should therefore be chosen. Because the lowpass filters are expected to suppress the upper sidebands that are located around twice the carrier frequency $2\,\omega_1$ (cf. Fig. 3.1a), ω_3 must be chosen to be significantly less than $2\,\omega_1$.

Last an equation for the pull-in time T_P will be derived. Equations (3.20), (3.21b), and (3.25) describe the behavior of the three building blocks in Fig. 3.7 and enable us to compute the three variables $\overline{u_d}$, $\overline{u_f}$, and $\Delta\omega$. We only need to know the instantaneous $\Delta\omega$ versus time; hence, we eliminate $\overline{u_d}$ and $\overline{u_f}$ from Eqs. (3.21b) and (3.25) and obtain the differential equation

$$\frac{d}{dt}\Delta\omega\,\,\tau_1 + \frac{1}{\Delta\omega}\,\frac{K_0^2 K_d^2 K_H}{\pi^2}\cos(\varphi_{tot}) = 0 \tag{3.29}$$

This differential equation is nonlinear, but the variables $\Delta\omega$ and t can be separated, which leads to an explicit solution. Putting all terms containing $\Delta\omega$ to the left side and performing an integration, we get

$$\frac{\tau_1 \pi^2}{K_0^2 K_d^2 K_H} \int_{\Delta\omega_0}^{\Delta\omega_L} \frac{\Delta\omega}{\cos(\varphi_{tot})} \, d\Delta\omega = - \int_0^{T_P} dt \qquad (3.30)$$

The limits of integration are $\Delta\omega_0$ and $\Delta\omega_L$ on the left side, because the pull-in process starts with an initial frequency offset $\Delta\omega = \Delta\omega_0$ and ends when $\Delta\omega$ reaches the value $\Delta\omega_L$, which is the lock range. Following that instant a lock-in process will start. The integration limits on the right side are 0 and T_P, respectively, which means that the pull-in process has duration T_P, and after that interval (fast) lock-in process starts.

Performing the integration on the left imposes some considerable problems, when we remember that $\cos(\varphi_{tot})$ is given by

$$\cos(\varphi_{tot}) = \cos\left(-2 \, arctg \, \frac{\Delta\omega}{\omega_3} - \frac{\pi}{2} + arctg \, \frac{2 \, \Delta\omega}{\omega_C}\right)$$

Finding an explicit solution for the integral seems difficult if not impossible, but the cos term can be drastically simplified. When we plot $\cos(\varphi_{tot})$ versus $\Delta\omega$, we observe that within the range $\Delta\omega_L < \Delta\omega < \Delta\omega_0$ the term $\cos(\varphi_{tot})$ is an almost perfect straight line. Hence, we can replace $\cos(\varphi_{tot})$ by

$$\cos(\varphi_{tot}) \approx 1 - \frac{\Delta\omega}{\Delta\omega_P}$$

Inserting that substitution into Eq. (3.30) yields a rational function of $\Delta\omega$ on the left side, which is easily integrated. After some mathematical manipulations, we obtain for the pull-in time T_P

$$T_P = \frac{\Delta\omega_P \pi^2 \tau_1}{K_0^2 K_d^2 K_H} \left[\Delta\omega_P \ln \frac{\Delta\omega_P - \Delta\omega_L}{\Delta\omega_P - \Delta\omega_0} - \Delta\omega_0 + \Delta\omega_L\right] \qquad (3.31)$$

Making use of Eqs. (3.11) and (3.15), we have

$$K_H = \frac{\tau_2}{\tau_1}$$

$$\omega_n^2 = \frac{K_0 K_d}{\tau_1}$$

$$\zeta = \frac{\omega_n \tau_2}{2}$$

Using these substitutions, Eq. (3.31) can be rewritten as

$$T_P = \frac{\Delta\omega_P \, \pi^2}{2 \, \zeta \, \omega_n^3} \left[\Delta\omega_P \ln \frac{\Delta\omega_P - \Delta\omega_L}{\Delta\omega_P - \Delta\omega_0} - \Delta\omega_0 + \Delta\omega_L\right] \qquad (3.32)$$

This equation is valid for initial frequency offsets in the range

$$\Delta\omega_L < \Delta\omega_0 < \Delta\omega_P$$

For lower frequency offsets, a fast pull-in process will occur and Eq. (3.13) should be used.

3.5 Design Procedures for Conventional Costas Loop for BPSK

In this section, we will describe design procedures for Costas loops that will be used for demodulation of BPSK signals. Two case studies will be presented, one for the design of analog and one for the design of digital systems.

3.5.1 Case Study 1: Designing an Analog Costas Loop for BPSK

An analog Costas loop for BPSK shall be designed in this section. It is assumed that a binary signal is modulated onto a carrier. The carrier frequency is set to 400 kHz, i.e., the Costas loop will operate at a center frequency $\omega_0 = 2 \pi 400'000 = 2'512'000$ rad s^{-1}. The symbol rate is assumed to be $f_S = 100'000$ symbols/s. Now the parameters of the loop (such as time constants τ_1 and τ_2, corner frequencies ω_C and ω_3, and gain parameters K_0 and K_d) must be determined [Note that these parameters have been defined in Eqs. (3.4–3.6), and (3.13)].

The modulation amplitude m_1 is set $m_1 = 1$, cf. Eq. (3.1). According to Eq. (3.3), the phase detector gain is then $K_d = 1$. It has proven advantageous to determine the remaining parameters by using the open loop transfer function $G_{OL}(s)$ of the loop [2]. This is given by

$$G_{OL}(s) = \frac{K_0 \, K_d}{s} \frac{1 + s/\omega_C}{s \, \tau_1} \frac{1}{1 + s/\omega_3} \tag{3.33}$$

The magnitude $|G_{OL}(\omega)|$ (Bode diagram) is plotted in Fig. 3.8. The magnitude curve crosses the 0 dB line at the so-called transit frequency ω_T. It is common practice to choose ω_T to be about $(0.05 \ldots 0.1) * \omega_0$ [2]. Here we set $\omega_T = 0.1 \, \omega_0$, i.e. $\omega_T = 251'200$ rad s^{-1}. Furthermore, we set corner frequency $\omega_C = \omega_T$. When doing so, the slope of the asymptotic magnitude curve changes from -40 dB/decade to -20 dB/decade at $\omega = \omega_C$. Under this condition, the phase of $G_{OL}(\omega)$ is $-135°$ at ω_C. Consequently, the phase margin of the loop becomes $45°$ which provides sufficient stability. According to Eq. (3.5), τ_2 becomes 4 μs. Next corner frequency ω_3 will be determined. The corner frequency of the lowpass filter must be chosen such that the demodulated data signal (i.e., the output of the lowpass filter in the I branch) is recovered with high fidelity. To fulfill this requirement, ω_3 should be chosen as large as possible. On the other hand, the lowpass filter should suppress the double frequency component (here at about 800 kHz) sufficiently, which means that ω_3 should be markedly less than 2 ω_0. It is a good compromise to set corner frequency to twice the symbol rate, i.e., $\omega_3 = 2 * 2 \pi * 100'000 = 1'256'000$ rad s^{-1}. Last the remaining parameters τ_1 and K_0 must be chosen. They have to be specified such that the open loop gain becomes 1 at frequency $\omega = \omega_C$. According to Eq. (3.33), we can set

$$G_{OL}(\omega_C) = 1 \approx \frac{K_0 \, K_d}{\omega_c^2 \, \tau_1} \tag{3.34}$$

Because two parameters are still undetermined, one of those can be chosen arbitrarily; hence, we set $\tau_1 = 20$ µs. Finally from (3.34), we get $K_0 = 1'262'000$ s^{-1}.

The design of the Costas loop is completed now, and we can compute the most important loop parameters. For the natural frequency and damping factor, we get from Eq. (3.11)

$$\omega_n = 251'000 \text{ rad}/s \ (f_n = 40 \text{ kHz})$$
$$\zeta = 0.5$$

From (3.12), the lock range becomes

$$\Delta\omega_L = 125'000 \text{ rad s} \ (\Delta f_L = 20 \text{ kHz})$$

and from (3.13), the lock time becomes

$$T_L = 25 \text{ µs}$$

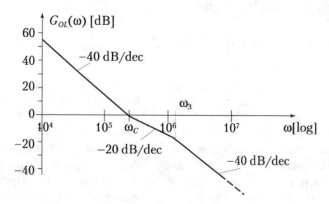

Fig. 3.8 Bode plot of open loop transfer function of Costas loop

Next we compute the pull-in range and get from Eq. (3.28)

$$\Delta\omega_P = 1'086'440 \text{ rad s}^{-1} \ (\Delta f_P = 173 \text{ kHz})$$

3.5.2 Case Study 2: Designing a Digital Costas Loop for BPSK

For the Costas loop, the same design parameters will be used as in the example before. To convert the analog loop into a digital one, we first must define a suitable sampling frequency f_{samp} (or sampling interval $T = 1/f_{samp}$). To satisfy the Nyquist theorem, the sampling frequency must be higher than twice the highest frequency that exists in the loop. In our case, the highest frequency is found at the output of the multipliers in the I and Q branches (cf. Fig. 3.1a). The sum frequency term is about twice the center frequency; hence, f_{samp} must be greater than four times the center frequency. A suitable choice would be $f_{samp} = 8 f_0 = 3.2$ MHz.

Next the transfer functions of the building block have to be converted into discrete transfer functions, i.e., $H(s) \rightarrow H(z)$. For best results, it is preferable to use the bilinear z transform [2]. Given an analog transfer function H(s), this can be converted into a discrete transfer function H(z) by replacing s by

$$s = \frac{2}{T}\frac{1-z^{-1}}{1+z^{-1}} \tag{3.35}$$

Now the bilinear z transform has the property that the analog frequency range from 0 to ∞ is compressed to the digital frequency range from 0 to $f_{samp}/2$. To avoid undesired "shrinking" of the corner frequencies (ω_C and ω_3), these must be "prewarped" accordingly, i.e., we must set

$$\omega_{C,p} = \frac{2}{T} tg\frac{\omega_C T}{2} \tag{3.36}$$

$$\omega_{3,p} = \frac{2}{T} tg\frac{\omega_3 T}{2} \tag{3.37}$$

where $\omega_{C,p}$ and $\omega_{3,p}$ are the prewarped corner frequencies. Now we can apply the bilinear z transform to the transfer functions of the lowpass filters [cf. Eq. (3.14)] and of the loop filter [cf. Eq. (3.5)] and get

$$H_{LPF}(z) = \frac{\left[1 + \frac{2}{\omega_{3,p} T}\right] + \left[1 - \frac{2}{\omega_{3,p} T}\right]z^{-1}}{1+z^{-1}} \tag{3.38}$$

$$H_{LF}(z) = \frac{\left[1 + \frac{2}{\omega_{c,p} T}\right] + \left[1 - \frac{2}{\omega_{c,p} T}\right] z^{-1}}{\frac{2\,\tau_1}{T} - \frac{2\,\tau_1}{T} z^{-1}} \tag{3.39}$$

Because the VCO is a simple integrator, we can apply the discrete z transform of an integrator, i.e.,

$$H_{VCO}(z) = \frac{K_0\,T}{1 - z^{-1}} \tag{3.40}$$

The digital Costas loop is ready now for implementation. A Simulink model will be presented in Sect. 3.6.

3.6 Simulating the Costas Loop for BPSK

The CD included with this book contains a number of Simulink models for different types of Costas loops. The models can be run directly from the CD but is generally more convenient to copy the files to hard disk. To do so, determine the folder where you want copy the files to; perhaps you will find it practical to copy them to the WORK folder found in every MATLAB installation. The models are stored in different subfolders. The model we used here is BPSK_Real.mdl and is in subfolder BPSK_Real. This model is designed following the procedure discussed in Sect. 3.6.

When the model is started, a dialog box is displayed, as shown in Fig. 3.9.

A number of parameters can be entered in the corresponding edit windows, such as carrier frequency, symbol rate. The button "Set initial defaults" can be used to set a number of initial values. Whenever one or more parameters have been changed, the "Done" button must be pressed to store that data.

The block diagram of the model is shown in Fig. 3.10. It does not only present the Costas loop but also the transmitting section, cf. the blocks on the left side.

The transmitter creates a random binary signal and modulates it to the carrier signal. All relevant signals are displayed by scopes. Scope labeled "RF Signals" shows the binary signal and the BPSK signal. Scope labeled I, Q shows the output signals if the I and Q branches. The I signal is equivalent to the binary signal c(t).

Detailed information is found in file "Description," which can be seen when clicking File/Model Properties/Description in the menu bar of the model.

It is instructive to play around with the parameters of the model, such as the receiver offset frequency for instance. When small frequency errors are specified, the loop will lock with the "correct" polarity. When choosing larger frequency, errors become possible that the loop locks with inverted polarity. It is also possible to measure the pull-in time for different values of initial frequency errors. Some results are shown in Table 3.1.

We note that the predicted and simulated parameters are in good agreement.

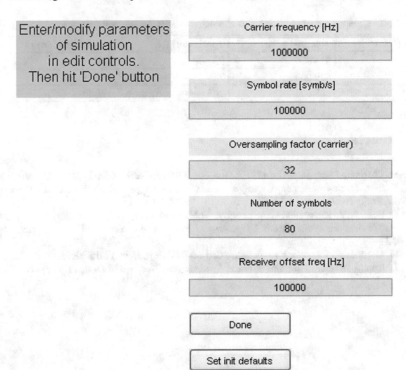

Fig. 3.9 Parameter window for Costas loop simulation

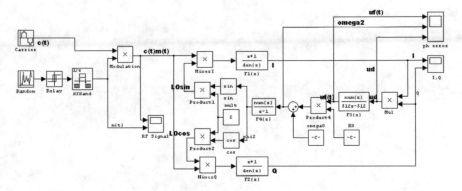

Fig. 3.10 Simulink model of the Costas loop for BPSK

Table 3.1 Comparison of predicted and simulated results for the pull-in range

Δf_0 (kHz)	$\Delta\omega_0$ (rad s^{-1})	T_P (theory) (µs)	T_P (simulation) (µs)
50	314′000	33	30
70	439′000	78	85
100	628′000	204	200

References

1. F.M. Gardner, *Phase-lock Techniques*, 2nd edn. (Wiley, New York, 1979)
2. R.E. Best, *Phase-locked Loops, Design, Simulation, and Applications*, 6th edn. (McGraw-Hill, New York, 2007)
3. R.E. Best, N.V. Kuznetsov, G.A. Leonov, M.V. Yuldashev, R.V. Yuldashev, in *Tutorial on Dynamic Analysis of the Costas Loop*, Annual Reviews in Control, vol. 42 (Elsevier, 2016), pp. 27–49
4. F.M. Gardner, in *Phaselock Techniques*, 3rd edn. (Wiley, New York, 2005)
5. U.L. Rohde, in *Microwave and Wireless Synthesizers, Theory and Design* (Wiley, New York, 1997)

Chapter 4
Conventional Costas Loop for QPSK

4.1 Linear Model and Frequency Response

The block diagram of the Costas loop for QPSK is shown in Fig. 4.1. The input signal $u_1(t)$ is defined as

$$u_1(t) = m_1 \cos(\omega_1 t + \theta_1) + m_2 \sin(\omega_1 t + \theta_1) \qquad (4.1)$$

m_1 and m_2 are the amplitudes of the modulating signals I and Q, respectively. The I signal is modulated onto the cosine carrier, and the Q signal is modulated onto the sine carrier. Let $m = |m_1| = |m_2|$. In many cases m is chosen 1, but it can have any other value. We first analyze the static behavior of the Costas loop. Both frequencies ω_1 and ω_2 are set equal, and the phase error $\theta_e = \theta_1 - \theta_2$ is varied over the range $0 < \theta_e < 2\pi$. According to Gardner [1], we then get for u_d

$$u_d = \begin{cases} 2m \sin \theta_e, & -\pi/4 < \theta_e < \pi/4 \\ -2m \cos \theta_e, & \pi/4 < \theta_e < 3\pi/4 \\ -2m \sin \theta_e, & 3\pi/4 < \theta_e < 5\pi/4 \\ 2m \cos \theta_e, & 5\pi/4 < \theta_e < 7\pi/4 \end{cases} \qquad (4.2)$$

u_d versus θ_e is plotted in Fig. 4.2. The curve looks like a "chopped" sine wave. The Costas loop can get locked at four different values of θ_e, i.e., with $\theta_e = 0$, $\pi/2$, π, or $3\pi/2$. To simplify the following analysis, we define the phase error to be zero wherever the loop gets locked. Moreover, in the locked state, the phase error is small, so we can write

---

Electronic supplementary material The online version of this chapter (doi:10.1007/978-3-319-72008-1_4) contains supplementary material, which is available to authorized users.

R. Best, *Costas Loops*, https://doi.org/10.1007/978-3-319-72008-1_4

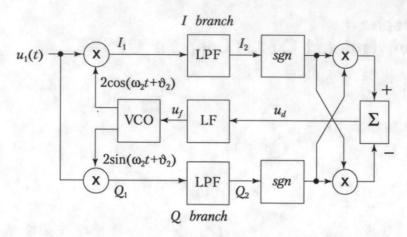

Fig. 4.1 Block diagram of Costas loop for QPSK

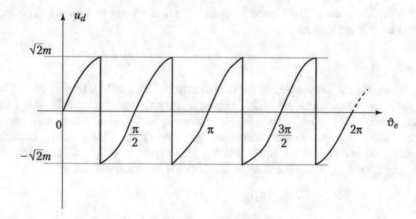

Fig. 4.2 Phase detector output signal u_d as a function of phase error θ_e

$$u_d \approx 2m\,\theta_e = K_d\,\theta_e \tag{4.3}$$

i.e., the output signal of the adder block at the right in Fig. 4.1 is considered to be the phase detector output signal u_d. The phase detector gain is then

$$K_d = 2m \tag{4.4}$$

It is easily seen that the linear model for the locked state is identical with that of the Costas loop for BPSK, cf. Fig. 3.1b. Because only small frequency differences are considered here, the lowpass filters in Fig. 4.1 can be discarded. The transfer functions of the loop filter and of the VCO are assumed to be the same as in case of the Costas loop for BPSK, and hence, these are given by Eqs. (3.5) and (3.6). The open loop transfer function is also identical with that of the Costas loop for BPSK, cf. Eq. (3.7) and Fig. 3.2. This holds true for the closed loop transfer function, too, cf. Eqs. (3.9), (3.10), and (3.11).

Next the lock range $\Delta\omega_L$ and the lock time T_L will be determined.

4.2   Lock Range $\Delta\omega_L$ and Lock Time T_L

To determine the lock range, we assume that the loop is out of lock. Let the reference frequency be ω_1 and the initial VCO frequency be ω_{20}. The difference frequency $\omega_1 - \omega_{20}$ is called $\Delta\omega$. When the loop has not acquired lock, the phase error θ_e is a continuously rising function that increases toward infinity. The phase detector output signal u_d is then a chopped sine wave as depicted in Fig. 4.2. The fundamental frequency of this signal is four times the difference frequency, i.e., $4\Delta\omega$. This signal is plotted once again in the top trace of Fig. 4.3. The amplitude of this signal is $K_d/\sqrt{2}$. The fundamental frequency of u_d is assumed to be much higher than the corner frequency ω_C of the loop filter; hence, the transfer function of the loop filter can be approximated by

$$H_{LF}(s) \approx \frac{\tau_2}{\tau_1} = K_H \qquad (4.5)$$

Hence, the output signal of the loop filter u_f has an amplitude of $K_d\,K_H/\sqrt{2}$, cf. middle trace of Fig. 4.3. This signal modulates the output frequency of the VCO, and the modulation amplitude is given by $K_d\,K_H\,K_0/\sqrt{2}$, cf. bottom trace in Fig. 4.3. It is easily seen that the loop spontaneously locks when the peak of the $\omega_2(t)$ waveform touches the ω_1 line; hence, we have

$$\Delta\omega_L = \frac{K_0 K_d K_H}{\sqrt{2}} \qquad (4.6)$$

Making use of Eqs. (3.11) and (4.5), this can be rewritten as

$$\Delta\omega_L = \sqrt{2}\zeta\omega_n \qquad (4.7)$$

Because the transient response of the loop is a damped oscillation whose frequency is ω_n, the loop will lock in at most one cycle of ω_n, and we get for the lock time

$$T_L \approx \frac{2\pi}{\omega_n}. \qquad (4.8)$$

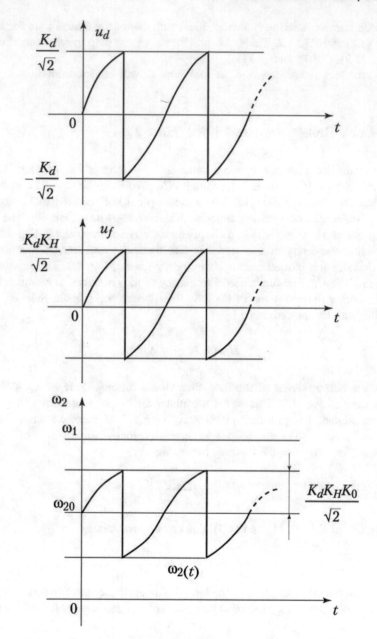

Fig. 4.3 Signals $u_d(t)$, $u_f(t)$, and $\omega_2(t)$ during the pull-in process

4.3   Nonlinear Model in the Unlocked State

The nonlinear model of the Costas loop for QPSK is developed on the basis of the nonlinear model we derived for the Costas loop for BPSK, cf. Sect. 3.3 and Fig. 3.4c [2–4]. Here again the order of lowpass filters and the blocks shown at the right of Fig. 4.1 is reversed. This results in the model shown in Fig. 4.4a. In the block labeled "B", the function blocks at the right of Fig. 4.4a have been integrated, cf. Fig. 4.4b. The output signal u_d of block B is the "chopped" sine wave as shown in Fig. 4.2. Its fundamental frequency is 4 times the frequency difference $\omega_1 - \omega_2$. The lowpass filters and the loop filter have been concatenated in the block labeled "LPF + LF" at the right of Fig. 4.4a. Referring to Fig. 4.1, signals I_1 and Q_1 are passed through lowpass filters. As in the case of the Costas loop for BPSK, we assume here again that the difference frequency $\Delta\omega$ is well below the corner frequency ω_3 of the lowpass filters, hence the gain of the lowpass filters is nearly 1 at $\omega = \Delta\omega$. Because the phase shift must not be neglected, we represent the lowpass filter by a delay, i.e., its frequency response at $\omega = \Delta\omega$ is

$$H_{LPF}(\Delta\omega) = \exp(j\varphi_1)$$

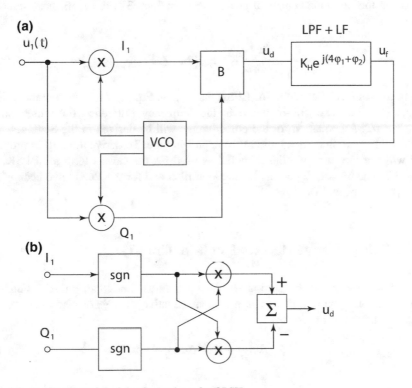

Fig. 4.4 Nonlinear model of the Costas loop for QPSK

where φ_1 is the phase of the lowpass filter at $\omega = \Delta\omega$. Due to the arithmetic operations in block "B" (cf. Fig. 4.4), the frequency of the u_d is quadrupled, which implies that the phase shift at frequency $4\Delta\omega$ becomes $4\varphi_1$. As for the Costas loop for BPSK, the frequency response of the loop filter at $\omega = 4\Delta\omega$ is given by

$$H_{LF}(4\,\Delta\omega) = K_H \exp(j\,\varphi_2)$$

where φ_2 is the phase of the loop filter at frequency $\omega = 4\Delta\omega$. K_H is the gain of the loop filter at "higher frequencies," cf. Eq. (3.15). Hence, the cascade of lowpass filter and loop filter can be modeled the transfer function $K_H\,exp(j[4\varphi_1 + \varphi_2])$ as shown in Fig. 4.4a. Let us define the total phase by $\varphi_{tot} = 4\varphi_1 + \varphi_2$.

Next we are computing the average phase detector output signal $\overline{u_d}$ as a function of frequency difference and phase φ_{tot}. First we calculate $\overline{u_d}$ for the special case $\varphi_{tot} = 0$. As shown in the bottom trace in Fig. 4.5 during interval T_1 the average frequency ω_2 is increased, hence the average difference $\Delta\omega$ becomes smaller. During next half cycle T_2, the reverse is true: The average difference $\Delta\omega$ becomes greater, hence for $\varphi_{tot} = 0T_1$ is longer than T_2. The modulating signal is therefore asymmetric, and because also $u_d(t)$ (top trace) is asymmetrical, its average $\overline{u_d}$ becomes non zero and positive. This asymmetry has been shown exaggerated in Fig. 4.5.

Using the same mathematical procedure as in Sect. 3.3, the average u_d signal is given by

$$\overline{u_d} = \frac{0.373^2 K_d^2 K_0 K_H}{\Delta\omega}\cos(4\varphi_1[\Delta\omega] + \varphi_2[4\Delta\omega]) \qquad (4.9)$$

As in case of the Costas loop for BPSK, here again Eq. (4.9) tells us that the pull-in range is finite. The pull-in range is the frequency difference for which phase $\varphi_{tot} = -\pi/2$. An equation for the pull-in range will be derived in the Sect. 4.4. We also will have to find an equation for the pull-in time. To derive the pull-in process, we will use the same nonlinear model as used for the Costas loop for BPSK, cf. Fig. 3.7. The transfer functions for the loop filter and for the VCO have been given in Eqs. (3.21b) and (3.25).

4.4   Pull-in Range $\Delta\omega_P$ and Pull-in Time T_P

The pull-in range can be computed using Eq. (4.9). Lock can only be obtained when the total phase shift φ_{tot} is not more negative than $-\pi/2$. This leads to an equation of the form

$$4\varphi_1(\Delta\omega_P) + \varphi_2(4\Delta\omega_P) = -\pi/2 \qquad (4.10)$$

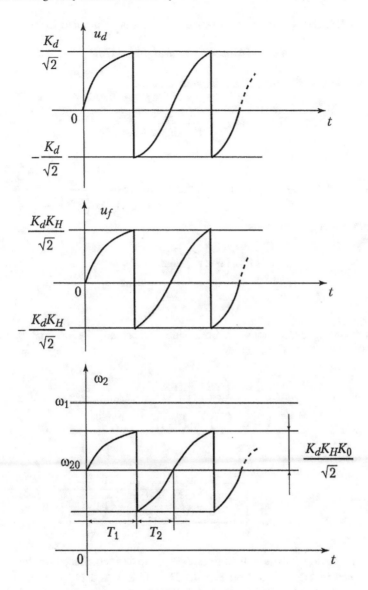

Fig. 4.5 Signals of the Costas loop for QPSK in the unlocked state

According to Eqs. (3.5) and (3.14), φ_1 and φ_2 are given by

$$\varphi_1(\omega) = -arctg(\omega/\omega_3), \quad \varphi_2(\omega) = -\pi/2 + arctg(\omega/\omega_C)$$

with $\omega_C = 1/\tau_2$. Hence, the pull-in range $\Delta\omega_P$ can be computed from

$$4\, arctg(\Delta\omega_P/\omega_3) = arctg(4\Delta\omega_P/\omega_C) \tag{4.11}$$

Using the addition theorem of the tangent function

$$tg(4\alpha) = \frac{(1 - tg^2\alpha)4\, tg\alpha}{1 - 6tg^2\alpha + tg^4\alpha}$$

the term $4\, arctg(\Delta\omega_P/\omega_3)$ can be replaced by $arc\, tg\, \dfrac{\left[1 - \left(\frac{\Delta\omega_P}{\omega_3}\right)^2\right]4\frac{\omega_P}{\omega_3}}{1 - 6\left(\frac{\Delta\omega_P}{\omega_3}\right)^2 + \left(\frac{\Delta\omega_P}{\omega_3}\right)^4}$

Equation (4.11) then reads

$$arc\, tg\, \frac{\left[1 - \left(\frac{\Delta\omega_P}{\omega_3}\right)^2\right]4\frac{\Delta\omega_P}{\omega_3}}{1 - 6\left(\frac{\Delta\omega_P}{\omega_3}\right)^2 + \left(\frac{\Delta\omega_P}{\omega_3}\right)^4} = arc\, tg\, \frac{4\Delta\omega_P}{\omega_C}$$

When the arc tg expressions on both sides are equal, the arguments must be identical as well, hence we get

$$\frac{\left[1 - \left(\frac{\Delta\omega_P}{\omega_3}\right)^2\right]4\frac{\Delta\omega_P}{\omega_3}}{1 - 6\left(\frac{\Delta\omega_P}{\omega_3}\right)^2 + \left(\frac{\Delta\omega_P}{\omega_3}\right)^4} = \frac{4\Delta\omega_P}{\omega_C}$$

Solving for $\Delta\omega_P$ yields

$$\Delta\omega_P = \omega_3\sqrt{\frac{6 - \frac{\omega_c}{\omega_3} - \sqrt{\left[6 - \frac{\omega_c}{\omega_3}\right]^2 - 4\left(1 - \frac{\omega_c}{\omega_3}\right)}}{2}} \tag{4.12}$$

Last an equation for the pull-in time T_P will be derived. Based on the nonlinear model shown in Fig. 3.7 and on Eqs. (3.21b), (3.25), and (4.9), we can create a differential equation for the instantaneous difference frequency $\Delta\omega$ as a function of time. For this type of Costas loop, the differential equation has the form

$$\frac{d}{dt}\Delta\omega\, \tau_1 + \frac{\cos(\varphi_{tot})}{\Delta\omega}0.373^2 K_0^2 K_d^2 K_H = 0$$

with

$$\cos(\varphi_{tot}) = -4 \; arctg \frac{\Delta\omega}{\omega_3} - \frac{\pi}{2} + arctg \frac{\Delta\omega}{\omega_C}$$

Also here the cos term can be replaced by

$$\cos(\varphi_{tot}) \approx 1 - \frac{\Delta\omega}{\Delta\omega_P}$$

and using similar procedures as in Sect. 3.4 we get for the pull-in time

$$T_P \approx \frac{\Delta\omega_P}{0.278\zeta \, \omega_n^3} \left[\Delta\omega_P \ln \frac{\Delta\omega_P - \Delta\omega_L}{\Delta\omega_P - \Delta\omega_0} - \Delta\omega_0 + \Delta\omega_L \right] \quad . \qquad (4.13)$$

which again is valid for initial frequency offsets in the range

$$\Delta\omega_L < \Delta\omega_0 < \Delta\omega_P$$

For lower frequency offsets, a fast pull-in process will occur, and Eq. (4.8) should be used.

4.5 Design Procedure for Costas Loop for QPSK

Case Study: Designing a digital Costas loop for QPSK

A digital Costas loop for QPSK shall be designed in this section. It is assumed that two binary signals (I and Q) are modulated onto a quadrature carrier (cosine and sine carrier). The carrier frequency is set to 400 kHz; i.e., the Costas loop will operate at a center frequency $\omega_0 = 2\pi \, 400'000 = 2'512'000$ rad s^{-1}. The symbol rate is assumed to be $f_S = 100'000$ symbols/s. Now the parameters of the loop (such as time constants τ_1 and τ_2, corner frequencies ω_C and ω_3, and gain parameters such as K_0, K_d) must be determined [Note that these parameters have been defined in Eqs. (3.4), (3.5), (3.6), and (3.13)].

The modulation amplitudes m_1 and m_2 are set to 1. According to Eq. (4.4), the phase detector gain is then $K_d = 2$. It has proven advantageous to determine the remaining parameters by using the open loop transfer function $G_{OL}(s)$ of the loop [5]. As shown in Sect. 3.5 [cf. Eq. (3.33)], this is given by

$$G_{OL}(s) = \frac{K_0 K_d}{s} \frac{1 + s/\omega_C}{s\tau_1} \frac{1}{1 + s/\omega_3} \qquad (4.14)$$

The magnitude of $G_{OL}(\omega)$ is shown in Fig. 4.6.

As already explained in Sect. 3.5, the magnitude curve crosses the 0 dB line at the transit frequency ω_T. As in the case of the Costas loop for BPSK, we again set

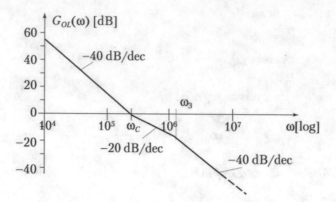

Fig. 4.6 Bode plot of open loop transfer function of Costas loop

$\omega_T = 0.1\omega_0$, i.e., $\omega_T = 251'200$ rad s^{-1}. Furthermore, we set corner frequency $\omega_C = \omega_T$. When doing so, the slope of the asymptotic magnitude curve changes from -40 dB/decade to -20 dB/decade at $\omega = \omega_C$. Under this condition, the phase of $G_{OL}(\omega)$ is $-135°$ at ω_C. Consequently, the phase margin of the loop becomes $45°$ which provides sufficient stability. According to Eq. (3.5), τ_2 becomes 4 μs. Next corner frequency ω_3 will be determined. The corner frequency of the lowpass filter must be chosen such that the demodulated data signal (i.e., the output of the lowpass filter in the I branch) is recovered with high fidelity. To fulfill this requirement, ω_3 should be chosen as large as possible. On the other hand, the lowpass filter should suppress the double frequency component (here at about 800 kHz) sufficiently, which means that ω_3 should be markedly less than $2\omega_0$. It's a good compromise to set corner frequency to twice the symbol rate, i.e., $\omega_3 = 2 * 2\pi * 100'000 = 1'256'000$ rad s^{-1}. Last the remaining parameters τ_1 and K_0 must be chosen. They have to be specified such that the open loop gain becomes 1 at frequency $\omega = \omega_C$. According to Eq. (4.14), we can set

$$G_{OL}(\omega_C) = 1 \approx \frac{K_0 K_d}{\omega_c^2 \tau_1} \qquad (4.15)$$

Because 2 parameters are still undetermined, one of those can be chosen arbitrarily, hence we set $\tau_1 = 20$ μs. Finally from (4.15), we get $K_0 = 631'000$ s^{-1}.

The design of the Costas loop is completed now, and we can compute the most important loop parameters. For the natural frequency and damping factor, we get from Eq. (3.11)

$$\omega_n = 251'000 \text{ rad/s } (f_n = 40 \text{ kHz})$$

$$\zeta = 0.5$$

From (4.7), the lock range becomes

$$\Delta\omega_L = 177'483 \text{ rad s}(\Delta f_L = 20 \text{ kHz})$$

and from (4.8) the lock time becomes

$$T_L = 25 \,\mu s$$

Next we want to compute the pull-in range. Solving (4.11) graphically yields

$$\Delta\omega_P = 439'600 \text{ rad s}^{-1}(\Delta f_P \approx 70 \text{ kHz})$$

Using the approximation (4.12), we get $\Delta\omega_P = 339'000$ rad s^{-1} ($\Delta f_P = 54$ kHz).

In Sect. 4.6, we will simulate this Costas loop and compare the results of the simulation with the predicted ones.

All block parameters have been determined now in the complex s domain. To get a digital Costas loop, we must convert the transfer function in the s domain into transfer functions in the z domain, using the z transform. As done in Sect. 3.6 a suitable sampling frequency f_{samp} must be chosen. As shown previously f_{samp} must be greater than 4 times the center frequency of the Costas loop. A suitable choice would be $f_{samp} = 8f_0 = 3.2$ MHz.

Next the transfer functions of the building block have to be converted into discrete transfer functions, i.e., $H(s) \rightarrow H(z)$. For best results, it is preferable to use the bilinear z transform [5]. Given an analog transfer function H(s), this can be converted into a discrete transfer function H(z) by replacing s by

$$s = \frac{2}{T}\frac{1-z^{-1}}{1+z^{-1}}$$

Now the bilinear z transform has the property that the analog frequency ranges from 0 to ∞ is compressed to the digital frequency range from 0 to $f_{samp}/2$. To avoid undesired "shrinking" of the corner frequencies (ω_C and ω_3), these must be "pre-warped" accordingly, i.e., we must set

$$\omega_{C,p} = \frac{2}{T} tg \frac{\omega_C T}{2}$$

$$\omega_{3,p} = \frac{2}{T} tg \frac{\omega_3 T}{2}$$

where $\omega_{C,p}$ and $\omega_{3,p}$ are the prewarped corner frequencies. Now we can apply the bilinear z transform to the transfer functions of the lowpass filters [cf. Eq. (3.14)] and of the loop filter [cf. Eq. (3.5)] and get

$$H_{LPF}(z) = \frac{\left[1 + \frac{2}{\omega_{3,p}T}\right] + \left[1 - \frac{2}{\omega_{3,p}T}\right]z^{-1}}{1 + z^{-1}}$$

$$H_{LF}(z) = \frac{\left[1 + \frac{2}{\omega_{C,p}T}\right] + \left[1 - \frac{2}{\omega_{C,p}T}\right]z^{-1}}{\frac{2\tau_1}{T} - \frac{2\tau_1}{T}z^{-1}}$$

Because the VCO is a simple integrator, we can apply the discrete z transform of an integrator, i.e.,

$$H_{VCO}(z) = \frac{K_0 T}{1 - z^{-1}}$$

The digital Costas loop is ready now for implementation. A Simulink model will be presented in Sect. 4.6.

4.6 Simulating Costas Loops for QPSK

To simulate a conventional Costas loop for QPSK, we use the model QPSK_Real. mdl. Its block diagram is shown in Fig. 4.7.

Two data signals (I and Q) are created by random number generators at the left in the block diagram. The other blocks are self-explanatory. The model is used now to check the validity of the approximations found for pull-in range and pull-in time.

Equation (4.13) predicted a theoretical pull-in range of $\Delta f_P = 73$ kHz. The simulations revealed a pull-in range of $\Delta f_P = 62$ kHz; hence, the prediction comes close to reality. A series of other simulations delivered results for the pull-in time ΔT_P. The results are listed in Table 4.1.

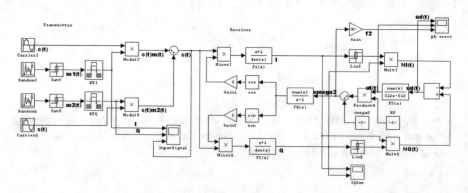

Fig. 4.7 Simulink model of the Costas loop for QPSK

Table 4.1 Comparison of predicted and simulated results for the pull-in range

Δf_0 (Hz)	$\Delta\omega_0$ (rad s^{-1})	T_P (theory) (μs)	T_P (simulation) (μs)
40 kHz	251'200	14	35
50 kHz	314'000	37	40
60 kHz	376'800	86	70

At higher frequency offsets, the results of the simulation are in good agreement with the predicted ones. The pull-in time for an initial frequency offset of 40 kHz is too low, however, but it should be noted that the lock time T_L is about 25 μs, and the total pull-in time cannot be less than the lock time.

References

1. M. Floyd, *Gardner, Phase-lock Techniques*, 2nd edn. (Wiley, New York, 1979)
2. R.E. Best, N.V. Kuznetsov, G.A. Leonow, M.V. Yuldashev, R.V. Yuldashev, Simulation of analog Costas loop circuits. Int. J. Automat. Comput. **11**, 571–579 (2014)
3. R.E. Best, N.V. Kuznetsov, G.A. Leonov, M.V. Yuldashev, R.V. Yuldashev, A short survey on nonlinear models of the classic Costas loop: rigorous derivation and limitation of the classic analysis, in *2015 American Control Conference*, Chicago, IL, USA, pp. 1296–1302, 1–3 July 2015
4. R.E. Best, N.V. Kuznetsov, G.A. Leonov, M.V. Yuldashev, R.V. Yuldashev, Tutorial on dynamic analysis of the Costas loop. Ann. Rev. Control **42**, Elsevier, 27–49 (2016)
5. E. Roland, *Best, Phase-locked Loops, Design, Simulation, and Applications*, 6th edn. (McGraw-Hill, New York, 2007)

Chapter 5
Modified Costas Loop for BPSK

The modified Costas loop for BPSK operates with so-called pre-envelope signals [1]. Figure 5.1a explains how a pre-envelope signal is generated.

A real input signal $u_1(t)$ is applied to the input of a Hilbert transformer [1, 2]. The output of the Hilbert transformer $\hat{u}_1(t)$ is considered to be the imaginary part of the pre-envelope signal, i.e., the pre-envelope signal is obtained from

$$u_1^+(t) = u_1(t) + j\hat{u}_1(t)$$

The transfer function $H_{hilb}(\omega)$ of the Hilbert transformer is defined by [3]

$$H_{hilb}(\omega) = \begin{vmatrix} -j, & \omega > 0 \\ j, & \omega < 0 \end{vmatrix}$$

i.e., the Hilbert transformer is a phase shifter. When a sine signal $\sin(\omega_1 t)$ is applied to the input of a Hilbert transformer, the output signal is shifted by 90° and becomes $\cos(\omega_1 t)$. When the input signal is a cosine signal $\cos(\omega_1 t)$, the output becomes— $\sin(\omega_1 t)$. The phase shift of 90° is identical for all frequencies. Hilbert transformers can be implemented by digital filters, generally by FIR filters [3, 4]. When the input signal of the Hilbert transformer is narrowband, it can be realized by a time delay, i.e., by delaying the signal by a quarter of one period of the carrier signal.

Electronic supplementary material The online version of this chapter (doi:10.1007/978-3-319-72008-1_5) contains supplementary material, which is available to authorized users.

R. Best, *Costas Loops*, https://doi.org/10.1007/978-3-319-72008-1_5

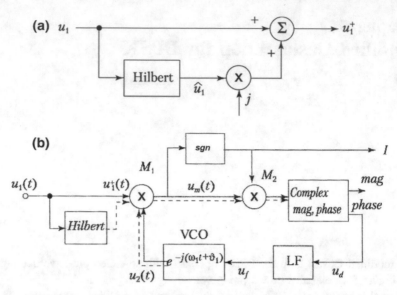

Fig. 5.1 **a** Generation of the pre-envelope signal using Hilbert transformer, **b** Block diagram of modified Costas loop for BPSK

5.1   Linear Model and Frequency Response

The block diagram of the modified Costas loop for BPSK is shown in Fig. 5.1b. Complex signals are represented by double lines; the real part of the signal is shown as a solid line and the imaginary part as a dashed line. The input signal is given by

$$u_1(t) = m_1 \cos(\omega_1 t + \theta_1)$$

with m_1 = modulating signal, ω_1 = reference frequency, and θ_1 = initial phase. m_1 can have two equal and opposite values, either +1 and −1, or +c and −c, where c can be an arbitrary constant. The input signal is first converted into a pre-envelope signal, as explained by the block diagram in Fig. 5.1a. The output signal of the Hilbert transformer (cf. Fig. 5.1b) is

$$\hat{u}_1(t) = H[m_1 \cos(\omega_1 t + \theta_1)] = m_1 \sin(\omega_1 t + \theta_1)$$

H(…) stands for Hilbert transform. (Note that because the largest frequency of the spectrum of the data signal m_1 is much lower than the carrier frequency ω_1, the Hilbert transform of the product $H[m_1 \cos(\omega_1 t + \theta_1)]$ equals $m_1 H[\cos(\omega_1 t + \theta_1)]$ [1].) The pre-envelope signal is obtained now from

$$u_1^+(t) = u_1(t) + j\hat{u}(t) = m_1 \exp(j[\omega_1 t + \theta_1]) \qquad (5.1)$$

The exponential in Eq. (5.1) is referred to as a "complex carrier." To demodulate the BPSK signal, the pre-envelope signal is now multiplied with the output signal of the VCO, which is here a complex carrier as well. The complex output signal of the VCO is defined as

$$u_2(t) = \exp(-j[\omega_2 t + \theta_2]) \qquad (5.2)$$

In the locked state of the Costas loop, both frequencies ω_1 and ω_2 are equal, and we also have $\theta_1 \approx \theta_2$. Hence, the output signal of the multiplier M_1 is

$$u_m(t) = m_1 \exp(j[(\omega_1 - \omega_2)t + \theta_1 - \theta_2]) \approx m_1 \qquad (5.3)$$

i.e., the output of the multiplier is the demodulated data signal m_1. To derive the linear model of this Costas loop, it is assumed that $\omega_1 = \omega_2$ and $\theta_1 \neq \theta_2$. The output signal of multiplier M_1 then becomes

$$u_m(t) = m_1 \exp(j[\theta_1 - \theta_2]) \qquad (5.4)$$

This is a phasor having magnitude $|m_1|$ and phase $\theta_1 - \theta_2$, as shown in Fig. 5.2. Two quantities are determined from the phase of phasor $u_m(t)$, i.e., the demodulated data signal I and the phase error θ_e. The data signal is defined as

$$I = \mathrm{sgn}(\mathrm{Re}[u_m(t)]) \qquad (5.5)$$

i.e., when the phasor lies in quadrants I or IV, the data signal is considered to be +1, and when the phasor is in quadrants II or III, the data signal is considered to be −1.

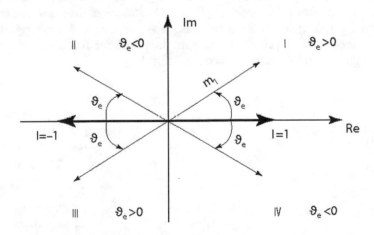

Fig. 5.2 Representation of phasor $u_m(t)$ in the complex plane

This means that I can be either a phasor with phase 0 or a phasor with phase π. These two phasors are plotted as thick lines as shown in Fig. 5.2.

The phase error θ_e is now given by the difference of the phases of phasor $u_m(t)$ and phasor I, as shown in Fig. 5.2, i.e., θ_e is determined from

$$\theta_e = phase(u_m(t) \cdot I) \tag{5.6}$$

The product $u_m(t) \cdot I$ is computed by multiplier M_2 in Fig. 5.2. The block labeled "Complex $\rightarrow$ mag, phase" is used to convert the complex signal delivered by M_2 into magnitude and phase. The magnitude is not used in this case, but only the phase. It follows from Eq. (5.6) that the phase output of this block is the phase error θ_e, hence the blocks M_1, M_2, sgn, and Complex $\rightarrow$ mag, phase represent a phase detector with gain $K_d = 1$. The phase output of block Complex $\rightarrow$ mag is therefore labeled u_d.

Figure 5.3 shows the completed linear model of the modified Costas loop for BPSK [5–7]. The transfer functions of the loop filter and VCO have been defined in Eqs. (3.5) and (3.6). Note that with this type of Costas loop there is no additional lowpass filter, because the multiplication of the two complex carriers [cf. Eq. (5.3)] does not create the unwanted double frequency component as found with the conventional Costas loops. From the model of Fig. 5.3, the open loop transfer function is determined to be

$$G_{OL}(s) = K_d \frac{K_0}{s} \frac{1 + s\tau_2}{s\tau_1}$$

Figure 5.4 shows a Bode plot of the magnitude of G_{OL}. The plot is characterized by the corner frequency ω_C which is defined by $\omega_C = 1/\tau_2$ and gain parameters K_d and K_0. At lower frequencies, the magnitude rolls off with a slope of -40 dB/decade. At frequency ω_C, the zero of the loop filter causes the magnitude to change its slope to -20 dB/decade. To get a stable system, the magnitude curve should cut the 0 dB line with a slope that is markedly less than -40 dB/decade. Setting the parameters such that the gain is just 0 dB at frequency ω_C provides a phase margin of 45° which assures stability [2]. From the open loop transfer function, we now can calculate the closed loop transfer function defined by

Fig. 5.3 Linear model of the modified Costas loop for BPSK

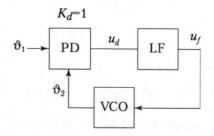

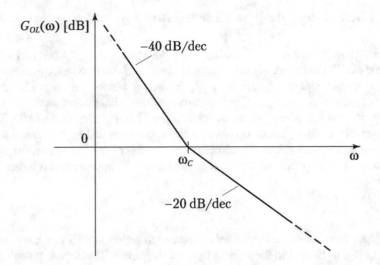

Fig. 5.4 Bode plot of magnitude of open loop gain $G_{OL}(\omega)$

$$G_{CL}(s) = \frac{\Theta_2(s)}{\Theta_1(s)}$$

After some mathematical manipulations, we get

$$G_{CL}(s) = \frac{K_0 K_d \frac{1+s\tau_2}{s\tau_1}}{s^2 + s\frac{K_0 K_d \tau_2}{\tau_1} + \frac{K_0 K_d}{\tau_1}}$$

It is customary to represent this transfer function in normalized form, i.e.,

$$G_{CS}(s) = \frac{2s\zeta\omega_n + \omega_n^2}{s^2 + 2s\zeta\omega_n + \omega_n^2}$$

with the substitutions

$$\omega_n = \sqrt{\frac{K_0 K_d}{\tau_1}} \quad , \quad \zeta = \frac{\omega_n \tau_2}{2} \tag{5.7}$$

where ω_n is called *natural frequency* and ζ is called *damping factor*. The linear model enables us to derive simple expressions for lock range $\Delta\omega_L$ and lock time T_L.

5.2 Lock Range $\Delta\omega_L$ and Lock Time T_L

For the following analysis, we assume that the loop is initially out of lock. The frequency of the reference signal (Fig. 5.1) is ω_1, and the frequency of the VCO is ω_2. The output signal of multiplier M_1 is then a phasor rotating with angular velocity $\Delta\omega = \omega_1 - \omega_2$. Consequently, the phase output of block "Complex $\rightarrow$ mag, phase" is a sawtooth signal having amplitude $(\pi/2)\,K_d$ and fundamental frequency $2\Delta\omega$, as shown in the upper trace of Fig. 5.5. Because $2\Delta\omega$ is usually much higher than the corner frequency ω_C of the loop filter, the transfer function of the loop filter at higher frequencies can be approximated again by

$$H_{LF}(\omega) \approx \frac{\tau_2}{\tau_1} = K_H$$

The output signal u_f of the loop filter is a sawtooth signal as well and has amplitude $(\pi/2)\,K_d\,K_H$, as shown in the middle trace of the figure. This signal modulates the frequency ω_2 generated by the VCO. The modulation amplitude is given by $(\pi/2)\,K_d\,K_H\,K_0$, cf. bottom trace. The Costas loop spontaneously acquires lock when the peak of the ω_2 waveform touches the ω_1 line, hence we have

Fig. 5.5 Signals u_d, u_f, and ω_2 during the lock process

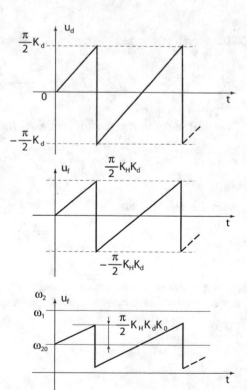

$$\Delta\omega_L = \frac{\pi}{2}K_d K_0 K_H = \frac{\pi}{2}K_d K_0 \frac{\tau_2}{\tau_1}$$

Making use of the substitutions Eq. (5.7), this can be rewritten as

$$\Delta\omega_L = \pi\,\zeta\,\omega_n \qquad\qquad\qquad (5.8)$$

Because the lock process is a damped oscillation having frequency ω_n, the lock time can be approximated by one cycle of this oscillation, i.e.,

$$T_L \approx \frac{2\pi}{\omega_n} \qquad\qquad\qquad (5.9)$$

5.3 Nonlinear Model for the Unlocked State

To derive the model for the unlocked state, we assume that the loop is not yet locked and that the difference between reference frequency ω_1 and VCO output frequency ω_2 is $\Delta\omega = \omega_1 - \omega_2$. As shown in Sect. 5.2 (cf. also Fig. 5.5), u_d is a sawtooth signal having frequency $2\Delta\omega$, cf. upper trace in Fig. 5.6. As will be explained in short, this signal is asymmetrical, i.e., the duration of the positive have wave T_1 is not identical with the duration T_2 of the negative half wave. The middle trace shows the output signal of the loop filter, and the lower trace shows the modulation of the VCO output frequency ω_2. From this waveform, it is seen that during T_1 the average frequency difference $\Delta\omega$ becomes smaller, but during interval T_2 it becomes larger. Consequently, the duration of T_1 is longer than the duration of T_2, and the average of signal u_d is nonzero and positive. Using the same mathematical procedure as in Sects. 3.3 and 4.3, the average $\overline{u_d}$ can be computed from [7]

$$\overline{u_d} = \frac{\pi^2 K_d K_0 K_H}{8\,\Delta\omega} \qquad\qquad\qquad (5.10)$$

Because this type of Costas loop does not require an additional lowpass filter, the u_d signal is not shifted in-phase, and therefore there is no cos term in Eq. (5.10). This implies that there is no polarity reversal in the function $\overline{u_d}(\Delta\omega)$, hence the pull-in range becomes theoretically infinite. Of course, in a real circuit, the pull-in range will be limited by the frequency range the VCO is capable to generate. When the center frequency f_0 of the loop is 10 MHz, for example, and when the VCO can create frequencies in the range from 0 to 20 MHz, then the maximum pull-in range Δf_P is 10 MHz, i.e., $\Delta\omega_P = 6.28 \times 10^6$ rad/s.

Figure 5.7 shows the nonlinear model used to compute the pull-in process. The block generating signal $\overline{u_d}$ is labeled "Phase/Frequency Detector" (cf. Eq. (5.10)) because in the unlocked state $\overline{u_d}$ us a function of the frequency error $\Delta\omega$.

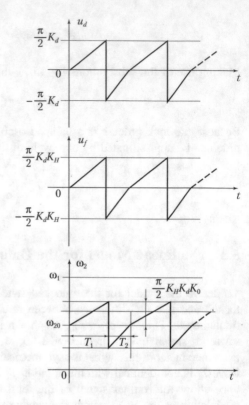

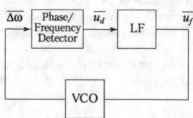

Using the same mathematical procedure as in Sect. 3.3, the transfer function of
the loop filter can be approximated as

$$H_{LF}(s) \approx \frac{1}{s\,\tau_1} \tag{5.11a}$$

In time domain, we can therefore write

$$\overline{u_f(t)} = \frac{1}{\tau_1} \int_0^t \overline{u_d(\tau)}\, d\tau \tag{5.11b}$$

For the transfer function of the VCO, we get

$$\Delta\omega = \Delta\omega_0 - K_0\overline{u_f} \qquad (5.12)$$

where $\Delta\omega_0$ is the initial frequency difference given by

$$\Delta\omega_0 = \omega_1 - \omega_{20}$$

Equations (5.10), (5.11b), and (5.12) enable us to compute the pull-in process, i.e., the instantaneous frequency difference $\Delta\omega$ as a function of time. This will be demonstrated in the next section.

5.4   Pull-in Range and Pull-in Time of the Modified Costas Loop for BPSK

As seen in the last section, the pull-in range of this type of Costas loop is theoretically infinite. This is, however, rather of "academic" interest. For real circuit "infinite pull-in range" simply means that the loop is able to lock onto frequencies that can be generated by the lock oscillator. Using Eqs. (5.10), (5.11b), and (5.12), we can derive an equation for the pull-in time [7]:

$$T_P \approx \frac{2}{\pi^2} \frac{\Delta\omega_0^2}{\zeta\,\omega_n^3} \qquad (5.13)$$

5.5   Design Procedure for Modified Costas Loop for BPSK

The following design is based on the method we already used in Sect. 3.5. It is assumed that a binary signal I is modulated onto a carrier. The carrier frequency is set to 400 kHz, i.e., the Costas loop will operate at a center frequency $\omega_0 = 2\pi$ 400'000 = 2'512'000 rad s^{-1}. The symbol rate is assumed to be $f_S = 100'000$ symbols/s. Now the parameters of the loop (such as time constants τ_1 and τ_2, corner frequency ω_C, and gain parameters such as K_0, K_d) must be determined [Note that these parameters have been defined in Eqs. (3.4), (3.5), (3.6), and (3.13)].

It has been shown in Sect. 5.1 that for this type of Costas loop $K_d = 1$. The modulation amplitudes m_1 and m_2 are set to 1. It has proven advantageous to determine the remaining parameters by using the open loop transfer function $G_{OL}(s)$ of the loop [2], which is given here by

$$G_{OL}(s) = K_d \frac{K_0}{s} \frac{1+s\tau_2}{s\tau_1}$$

The magnitude of $G_{OL}(\omega)$ has been shown in Fig. 5.4. As already explained in Sect. 3.5, the magnitude curve crosses the 0 dB line at the transit frequency ω_T. As in the case of the conventional Costas loop for BPSK, we again set $\omega_T = 0.1\omega_0$, i.e., $\omega_T = 251'200$ rad s^{-1}. Furthermore, we set corner frequency $\omega_C = \omega_T$. When doing so, the slope of the asymptotic magnitude curve changes from -40 dB/decade to -20 dB/decade at $\omega = \omega_C$. Under this condition, the phase of $G_{OL}(\omega)$ is $-135°$ at ω_C. Consequently, the phase margin of the loop becomes $45°$ which provides sufficient stability. According to Eq. (3.5), τ_2 becomes 4 µs. Last the remaining parameters τ_1 and K_0 must be chosen. They have to be specified such that the open loop gain becomes 1 at frequency $\omega = \omega_C$. According to Eq. (4.14), we can set

$$G_{OL}(\omega_C) = 1 \approx \frac{K_0\, K_d}{\omega_c^2\, \tau_1}$$

Because two parameters are still undetermined, one of those can be chosen arbitrarily, hence we set $\tau_1 = 20$ µs. We then get $K_0 = 1'262'000$ s^{-1}.

The design of the Costas loop is completed now, and we can compute the most important loop parameters. For the natural frequency and damping factor, we get from Eq. (3.11)

$$\omega_n = 251'000\, \text{rad/s}\ (f_n = 40\,\text{kHz})$$

$$\zeta = 0.5$$

From (5.7), the lock range becomes

$$\Delta\omega_L = 394'000\, \text{rad/s}\ (\Delta f_L = 62.7\,\text{kHz})$$

and from (5.8), the lock time becomes

$$T_L = 25\,\text{µs}$$

All block parameters have been determined now in the complex s domain. To get a digital Costas loop, we must convert the transfer function in the s domain into transfer functions in the z domain, using the z transform. As done in Sect. 3.6, a suitable sampling frequency f_{samp} must be chosen. As shown previously, f_{samp} must be greater than 4 times the center frequency of the Costas loop. A suitable choice would be $f_{samp} = 8f_0 = 3.2$ MHz.

Next the transfer functions of the building block have to be converted into discrete transfer functions, i.e., $H(s) \rightarrow H(z)$. For best results, it is preferable to use the bilinear z transform [2]. Given an analog transfer function H(s), this can be converted into a discrete transfer function H(z) by replacing s by

$$s = \frac{2}{T}\frac{1 - z^{-1}}{1 + z^{-1}}$$

Now the bilinear z transform has the property that the analog frequency ranges from 0 to ∞ is compressed to the digital frequency ranging from 0 to $f_{samp}/2$. To avoid undesired "shrinking" of the corner frequency ω_C, this must be "prewarped" accordingly, i.e., we must set

$$\omega_{C,p} = \frac{2}{T} tg \frac{\omega_C T}{2}$$

where $\omega_{C,p}$ is the prewarped corner frequency. Now we can apply the bilinear z transform to the transfer functions of the loop filter [cf. Eq. (3.5)] and get

$$H_{LF}(z) = \frac{\left[1 + \frac{2}{\omega_{C,p} T}\right] + \left[1 - \frac{2}{\omega_{C,p} T}\right] z^{-1}}{\frac{2 \tau_1}{T} - \frac{2 \tau_1}{T} z^{-1}}$$

Because the VCO is a simple integrator, we can apply the discrete z transform of an integrator, i.e.,

$$H_{VCO}(z) = \frac{K_0 T}{1 - z^{-1}}$$

The digital Costas loop is ready now for implementation. A Simulink model will be presented in Sect. 5.6.

5.6 Simulating Modified Costas Loops for BPSK

We will perform simulations for the modified Costas loop for BPSK on two different models:

1. Model **BPSK_Comp.mdl**
   A digital modified Costas loop for BPSK.
2. Model **BPSK_Comp_PreAmb.mdl**
   A digital modified Costas loop using a preamble that forces the loop to lock with "correct" polarity.

Model BPSK_Comp.mdl

The block diagram of the model is shown in Fig. 5.8.

The phase detector of this circuit operates as described in Sect. 5.1 and Eq. (5.6). When larger frequency errors (receiver offset frequency) are chosen in the dialog box of the model, the loop can lock with inverted polarity. The pull-in time has been determined for a number of different receiver offset frequencies. The results are shown in Table 5.1.

The predictions for $\Delta f_0 = 50$ kHz and 100 kHz are too low. As already mentioned in Sect. 4.7, the pull-in time cannot be lower than the lock time, and the

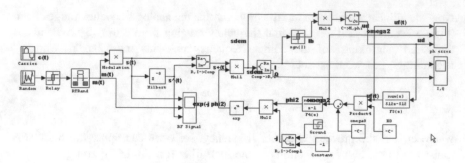

Fig. 5.8 Simulink model of the modified Costas loop for BPSK

Table 5.1 Comparison of predicted and simulated results for the pull-in range

Δf_0 (Hz)	$\Delta \omega_0$ (rad s^{-1})	T_P (theory) (µs)	T_P (simulation) (µs)
50 kHz	314'200	2.5	20
100 kHz	628'000	10	20
200 kHz	1'256'000	40	50

latter is estimated ≈25 µs. The simulation results for these two different frequencies are around 20 µs, which roughly correspond with the lock time. The simulation result for a frequency difference of 200 kHz comes close to the predicted value.

Model BPSK_Comp_PreAmb.mdl

The block diagram of this model is shown in Fig. 5.9. The circuit is almost identical with the previously discussed model (Fig. 5.8), but the phase detector is different (cf. the blocks sgn(I), Switch2, Mul4, C->M, phi). This phase detector operates in two different modes: (1) in "preamble mode" during the preamble interval and (2) in "data mode" after the preamble.

The transmitter generates a modulating signal labeled m(t). During the preamble interval m(t) is constant, i.e., m(t) = 1. After the preamble, the modulating signal is a random binary signal as in the previous model. The transmitter also contains a step function block labeled Step 1. During the preamble interval, the output signal of Step 1 is 1, and after the preamble, it is 0. This signal is used to switch the phase detector between the two modes mentioned above.

In preamble mode, the phase detector computes the phase error θ_e from θ_e = phase($u_m(t)$), cf. also Eq. (5.6). In the block diagram of Fig. 5.9, $u_m(t)$ is identical with the signal labeled sdem. In data mode, however, the phase error is computed from θ_e = phase(I $u_m(t)$), where I is the sign of the real part of $u_m(t)$. Let us consider the phase detector operation by a numerical example. Suppose that the current signal $u_m(t)$ is a phasor having a phase angle of 170°, as shown in Fig. 5.10. When the phase detector operates in preamble mode, it "sees" a phase error of 170°.

This forces the loop to rotate phasor $u_m(t)$ in clockwise direction until it approaches a position of approximately $u_m(t) = 1 + j \cdot 0$. In this way, it is

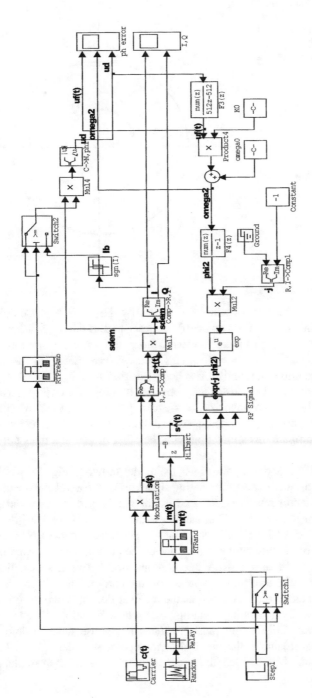

Fig. 5.9 Block diagram of modified Costas loop using preamble for synchronization

Fig. 5.10 Operation of the
phase detector in Fig. 5.9 in
preamble mode and data
mode, respectively

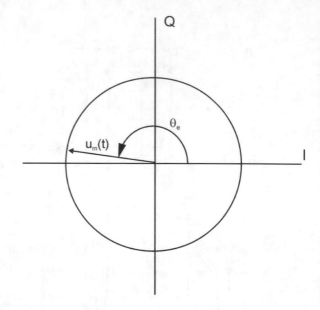

prevented that the loop locks with "false" polarity, i.e., with a phase difference of 180° between the signals $u_1(t)$ and $u_2(t)$, as described in Chap. 2. When the phase detector operates in data mode, however, the phase error becomes θ_e = phase(I $u_m(t)$) = phase$((-1) \cdot u_m(t))$ = $-10°$. This causes the phasor to be rotated in counterclockwise direction until it reaches a position near $u_m(t) = -1 + j \cdot 0$.

To see how this type of Costas loop operates it is instructive to run a simulation with a relatively large receiver offset frequency, e.g., 100 kHz. Monitoring the scopes RF Signal and I, Q will demonstrate the loop locks with correct polarity. It is interesting to compare this simulation with the other model that does not make use of a preamble.

To be useful in practical applications, the model shown in Fig. 5.9 would have to be extended in different ways. The gating signal used to control the operating mode of the phase detector must be generated by the receiver, of course, and cannot be delivered by the transmitter. This can be realized when data are transmitted in blocks having limited duration. Such a block could consist, e.g., of N preamble bits, i.e., by a series of N consecutive logical ones. The preamble is followed by a series of data bits, e.g., M data bits. A pause is inserted between succeeding blocks. During the pause, the transmitter switches off the carrier. The receiver is equipped with a carrier detect circuit. As soon as the receiver detects a carrier, it initiates the preamble interval by generating a gating pulse whose duration is identical with the preamble interval. This gating pulse switches the phase detector into preamble mode. After the preamble, the phase detector operates in data mode. Data transmission in blocks of limited length is very frequently used, for example, in mobile phones [8].

References

1. S.A. Tretter, *Communications System Design Using DSP Algorithms* (Springer Science + Business Media, 2008)
2. E. Roland, *Best, Phase-locked Loops, Design, Simulation, and Applications*, 6th edn. (McGraw-Hill, New York, 2007)
3. A.V. Oppenheim, R.W. Schafer, *Discrete-Time Signal Processing* (Prentice-Hall, US, 1989)
4. R.J. Higgins, *Digital Signal Processing in VLSI* (Prentice Hall, US, 1990)
5. R.E. Best, N.V. Kuznetsov, G.A. Leonow, M.V. Yuldashev, R.V. Yuldashev, Simulation of analog Costas loop circuits. Int. J. Automat. Comput. **11**, 571–579 (2014)
6. R.E. Best, N.V. Kuznetsov, G.A. Leonov, M.V. Yuldashev, R.V. Yuldashev, A short survey on nonlinear models of the classic Costas loop: rigorous derivation and limitation of the classic analysis, 2015 American Control Conference, Chicago, IL, USA, 1296–1302, 1–3 July 2015
7. R.E. Best, N.V. Kuznetsov, G.A. Leonov, M.V. Yuldashev, R.V. Yuldashev, Tutorial on dynamic analysis of the Costas loop. Ann. Rev. Control **42**, Elsevier, 27–49 (2016)
8. U.L. Rohde, D.P. Newkirk, *RF/Microwave Circuit Design for Wireless Applications* (Wiley, New York, 2000)

Chapter 6
Modified Costas Loop for QPSK

6.1 Operating Principle

Figure 6.1 shows the block diagram of the modified Costas loop for QPSK [1–4].
 The reference signal $u_1(t)$ is defined by

$$u_1(t) = m_1 \cos(\omega_1 t + \theta_1) - m_2 \sin(\omega_1 t + \theta_1) \qquad (6.1)$$

where m_1 and m_2 are data signals that can have a value of +c or −c, where c is an arbitrary constant. In many cases, c = 1. The Hilbert transformed signal is then given by

$$\hat{u}_1(t) = m_1 \sin(\omega_1 t + \theta_1) + m_2 \cos(\omega_1 t + \theta_1) \qquad (6.2)$$

and the pre-envelope signal then becomes

$$u_1^+(t) = m_1 \cos(\omega_1 t + \theta_1)$$
$$- m_2 \sin(\omega_1 t + \theta_1) + jm_1 \sin(\omega_1 t + \theta_1) + jm_2 \cos(\omega_1 t + \theta_1) \qquad (6.3)$$

This can be rewritten as

$$u_1^+(t) = (m_1 + jm_2)(\cos[\omega_1 t + \theta_1] + j \sin[\omega_1 t + \theta_1])$$
$$= (m_1 + jm_2) \exp(j[\omega_1 t + \theta_1]) \qquad (6.4)$$

Herein the term $m_1 + j\,m_2$ is called "complex envelope" [5], and the term $exp(j\omega_1 t + \theta_1)$ is referred to as "complex carrier." The VCO generates another complex carrier given by

---

Electronic supplementary material The online version of this chapter (doi:10.1007/978-3-319-72008-1_6) contains supplementary material, which is available to authorized users.

65
R. Best, *Costas Loops*, https://doi.org/10.1007/978-3-319-72008-1_6

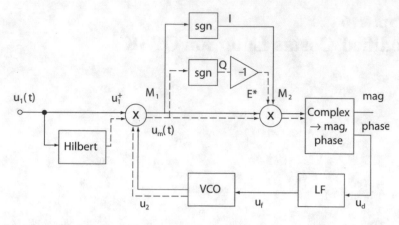

Fig. 6.1 Block diagram of modified Costas loop for QPSK

$$u_2(t) = \exp(-j[\omega_2 t + \theta_2]) \tag{6.5}$$

The multiplier M_1 creates signal $u_m(t)$ that is given by

$$u_m(t) = (m_1 + jm_2)\exp(j[(\omega_1 - \omega_2)t + (\theta_1 - \theta_2)]) \tag{6.6}$$

When the loop has acquired lock, $\omega_1 = \omega_2$ and $\theta_1 \approx \theta_2$, so we have

$$u_m(t) \approx (m_1 + jm_2) \tag{6.7}$$

Hence, the output of M_1 is the complex envelope. In the locked state, the complex envelope can take four positions, as shown in Fig. 6.2. When there is a phase error, $u_m(t)$ deviates from the ideal position, as demonstrated in the figure. The phase error θ_e is the angle between $u_m(t)$ and the closest of the four possible positions. When $u_m(t)$ is in quadrant I, e.g., phasor $1 + j$ is considered the estimate of the complex envelope. When $u_m(t)$ is in quadrant II, however, the estimate of the complex envelope would be $-1 + j$, etc. The estimates I and Q are taken from the output of sgn blocks, cf. Fig. 6.1. The phase error is obtained from

$$\theta_e = phase[u_m(t) \cdot (I - jQ)] \tag{6.8}$$

where $I - jQ$ is the conjugate of the complex envelope. Multiplier M_2 delivers the product $u_m(t) \cdot (I - jQ)$, and the block "Complex $\rightarrow$ mag, phase" is used to compute the phase of that complex quantity. Note that the magnitude is not required. As we already have seen in Sect. 6.1, the blocks M_1, sgn, Inverter, M_2, and Complex $\rightarrow$ mag, phase form a phase detector having gain $K_d = 1$. The phase output of block Complex $\rightarrow$ mag, phase is therefore labeled u_d.

Fig. 6.2 Representation of phasor $u_m(t)$ in the complex plane

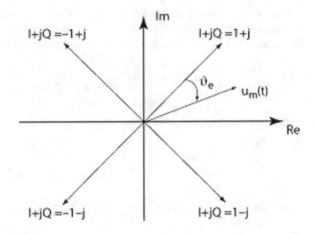

6.2   The Transfer Function of the Modified Costas Loop for QPSK

Figure 6.3 shows the completed linear model of the modified Costas loop for QPSK.

The transfer functions of the loop filter and VCO have been defined in Eqs. (3.5) and (3.6). From the model of Fig. 6.3, the open loop transfer function is determined to be

$$G_{OL}(s) = K_d \frac{K_0}{s} \frac{1 + s\tau_2}{s\tau_1}$$

Figure 6.4 shows a Bode plot of the magnitude of G_{OL}. The plot is characterized by the corner frequency ω_C which is defined by $\omega_C = 1/\tau_2$, and gain parameters K_d and K_0. At lower frequencies, the magnitude rolls off with a slope of -40 dB/decade. At frequency ω_C, the zero of the loop filter causes the magnitude to change its slope to -20 dB/decade. To get a stable system, the magnitude curve should cut the 0 dB line with a slope that is markedly less than -40 dB/decade. Setting the parameters

Fig. 6.3 Linear model of the Costas loop

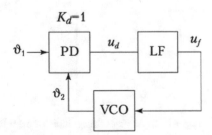

such that the gain is just 0 dB at frequency ω_C provides a phase margin of 45° which assures stability [6]. From the open loop transfer function, we now can calculate the closed loop transfer function defined by

$$G_{CL}(s) = \frac{\Theta_2(s)}{\Theta_1(s)}$$

After some mathematical manipulations, we get

$$G_{CL}(s) = \frac{K_0 K_d \frac{1+s\tau_2}{s\tau_1}}{s^2 + s\frac{K_0 K_d \tau_2}{\tau_1} + \frac{K_0 K_d}{\tau_1}}$$

It is customary to represent this transfer function in normalized form, i.e.,

$$G_{CS}(s) = \frac{2s\zeta\omega_n + \omega_n^2}{s^2 + 2s\zeta\omega_n + \omega_n^2}$$

with the substitutions

$$\omega_n = \sqrt{\frac{K_0 K_d}{\tau_1}}, \quad \zeta = \frac{\omega_n \tau_2}{2} \tag{6.9}$$

where ω_n is called *natural frequency* and ζ is called *damping factor*. The linear model enables us to derive simple expressions for lock range $\Delta\omega_L$ and lock time T_L.

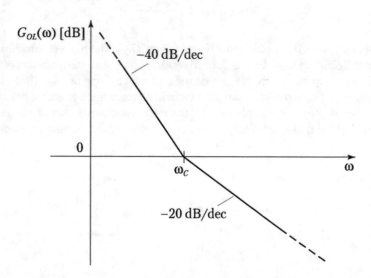

Fig. 6.4 Bode plot of magnitude of open loop gain $G_{OL}(\omega)$

6.3   Lock Range $\Delta\omega_L$ and Lock Time T_L

For the following analysis, we assume that the loop is initially out of lock. The frequency of the reference signal (Fig. 6.1) is ω_1, and the frequency of the VCO is ω_2. The output signal of multiplier M_1 is then a phasor rotating with angular velocity $\Delta\omega = \omega_1 - \omega_2$. Consequently, the phase output of block "Complex $\rightarrow$ mag, phase" is a sawtooth signal having amplitude $(\pi/4)\,K_d$ and fundamental frequency $4\,\Delta\omega$, as shown in the upper trace of Fig. 6.5. Because $4\,\Delta\omega$ is usually much higher than the corner frequency ω_C of the loop filter, the transfer function of the loop filter at higher frequencies can be approximated again by

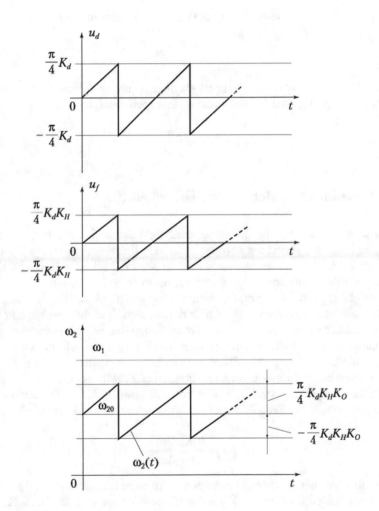

Fig. 6.5 Signals u_d, u_f, and ω_2 during the lock process

$$H_{LF}(\omega) \approx \frac{\tau_2}{\tau_1} = K_H$$

The output signal u_f of the loop filter is a sawtooth signal as well and has amplitude $(\pi/4) \, K_d \, K_H$, as shown in the middle trace of the figure. This signal modulates the frequency ω_2 generated by the VCO. The modulation amplitude is given by $(\pi/4)$ $K_d \, K_H \, K_0$, cf. bottom trace. The Costas loop spontaneously acquires lock when the peak of the ω_2 waveform touches the ω_1 line; hence, we have

$$\Delta\omega_L = \frac{\pi}{4} K_d K_0 K_H = \frac{\pi}{4} K_d K_0 \frac{\tau_2}{\tau_1} \tag{6.10}$$

Making use of the substitutions Eq. (5.7) this can be rewritten as

$$\Delta\omega_L = \frac{\pi}{2} \zeta \, \omega_n \tag{6.11}$$

Because the lock process is a damped oscillation having frequency ω_n, the lock time can be approximated by one cycle of this oscillation, i.e.,

$$T_L \approx \frac{2\pi}{\omega_n} \tag{6.12}$$

6.4   NonLinear Model for the Unlocked State

To derive the model for the unlocked state, we assume that the loop is not yet locked and that the difference between reference frequency ω_1 and VCO output frequency ω_2 is $\Delta\omega = \omega_1 - \omega_2$. As shown in Sect. 6.2 (cf. also Fig. 6.5), u_d is a sawtooth signal having frequency $4 \, \Delta\omega$, cf. upper trace in Fig. 6.6.

As will be explained in short, this signal is asymmetrical, i.e., the duration of the positive have wave T_1 is not identical with the duration T_2 of the negative half wave. The middle trace shows the output signal of the loop filter, and the lower trace shows the modulation of the VCO output frequency ω_2. From this waveform, it is seen that during T_1 the average frequency difference $\Delta\omega$ becomes smaller, but during interval T_2, it becomes larger. Consequently, the duration of T_1 is longer than the duration of T_2, and the average of signal u_d is nonzero and positive. Using the same mathematical procedure as in Sects. 3.3 and 4.3, the average $\overline{u_d}$ can be computed from

$$\overline{u_d} = \frac{\pi^2 K_d^2 K_0 K_H}{64 \, \Delta\omega} \tag{6.13}$$

Because this type of Costas loop does not require an additional lowpass filter, the u_d signal is not shifted in-phase, and therefore there is no cos term in Eq. (6.12). This

Fig. 6.6 Pull-in process of
the modified Costas loop for
QPSK

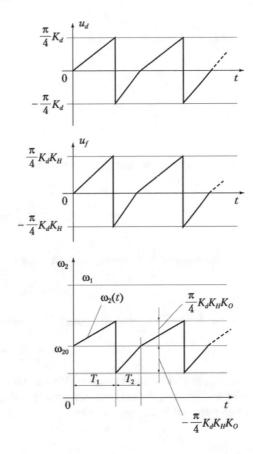

implies that there is no polarity reversal in the function $\overline{u_d}(\Delta\omega)$; hence, the pull-in range becomes theoretically infinite. Of course, in a real circuit the pull-in range will be limited by the frequency range the VCO is capable to generate. When the center frequency f_0 of the loop is 10 MHz, for example, and when the VCO can create frequencies in the range from 0 to 20 MHz, then the maximum pull-in range Δf_P is 10 MHz, i.e., $\Delta\omega_P = 6.28 \; 10^6$ rad/s.

Figure 6.7 shows the nonlinear model used to compute the pull-in process. The block generating signal $\overline{u_d}$ is labeled "Phase/Frequency Detector" [cf. Eq. (5.10)] because in the unlocked state $\overline{u_d}$ is a function of the frequency error $\Delta\omega$.

Using the same mathematical procedure as in Sect. 3.3, the transfer function of the loop filter can be approximated as

$$H_{LF}(s) \approx \frac{1}{s\,\tau_1} \tag{6.14}$$

Fig. 6.7 Nonlinear model of
the modified Costas loop

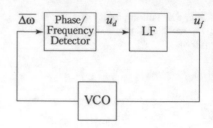

In time domain, we can therefore write

$$\overline{u_f(t)} = \frac{1}{\tau_1} \int_0^t \overline{u_d(\tau)} \, d\tau \qquad (6.15)$$

For the transfer function of the VCO, we get

$$\Delta\omega = \Delta\omega_0 - K_0\overline{u_f} \qquad (6.16)$$

where $\Delta\omega_0$ is the initial frequency difference given by

$$\Delta\omega_0 = \omega_1 - \omega_{20}$$

Equations (6.13), (6.14), and (6.15) enable us to compute the pull-in process, i.e., the instantaneous frequency difference $\Delta\omega$ as a function of time. This will be demonstrated in the next section.

6.5 Pull-in Range and Pull-in Time of the Modified Costas Loop for QPSK

As seen in the last section, the pull-in range of this type of Costas loop can be arbitrarily large. Using Eqs. (6.12), (6.13), and (6.14), we can derive an equation for the pull-in time T_P:

$$T_P \approx \frac{16}{\pi^2} \frac{\Delta\omega_0^2}{\zeta \, \omega_n^3} \qquad (6.17)$$

6.6 Design Procedure for Modified Costas Loop for QPSK

In the following, a design procedure for a digital modified Costas loop is presented. The design is based on the method and we already used in Sect. 5.5. It is assumed that two binary signals (I and Q) are modulated onto a quadrature carrier (cosine and sine carrier). The carrier frequency is set to 400 kHz, i.e., the Costas loop will operate at a center frequency $\omega_0 = 2 \pi 400'000 = 2'512'000$ rad s^{-1}. The symbol rate is assumed to be $f_S = 100'000$ symbols/s. Now the parameters of the loop (such as time constants τ_1 and τ_2, corner frequency ω_C, and gain parameters such as K_0, K_d) must be determined. [Note that these parameters have been defined in Eqs. (3.4), (3.5), (3.6), and (3.13)].

It has been shown in Sect. 6.1 that for this type of Costas loop $K_d = 1$. The modulation amplitudes m_1 and m_2 are set to 1. It has proven advantageous to determine the remaining parameters by using the open loop transfer function $G_{OL}(s)$ of the loop [6], which is given here by

$$G_{OL}(s) = K_d \frac{K_0}{s} \frac{1 + s\tau_2}{s\tau_1}$$

The magnitude of $G_{OL}(\omega)$ has been shown in Fig. 6.4. As already explained in Sect. 3.5, the magnitude curve crosses the 0 dB line at the transit frequency ω_T. As in the case of the Costas loop for BPSK, we again set $\omega_T = 0.1 \; \omega_0$, i.e., $\omega_T = 251'200$ rad s^{-1}. Furthermore, we set corner frequency $\omega_C = \omega_T$. When doing so, the slope of the asymptotic magnitude curve changes from -40 dB/decade to -20 dB/decade at $\omega = \omega_C$. Under this condition, the phase of $G_{OL}(\omega)$ is $-135°$ at ω_C. Consequently, the phase margin of the loop becomes $45°$ which provides sufficient stability. According to Eq. (3.5), τ_2 becomes 4 µs. Last the remaining parameters τ_1 and K_0 must be chosen. They have to be specified such that the open loop gain becomes 1 at frequency $\omega = \omega_C$. According to Eq. (5.14), we can set

$$G_{OL}(\omega_C) = 1 \approx \frac{K_0 \, K_d}{\omega_c^2 \, \tau_1}$$

Because 2 parameters are still undetermined, one of those can be chosen arbitrarily; hence, we set $\tau_1 = 20$ µs. We then get $K_0 = 1'262'000$ s^{-1}.

The design of the Costas loop is completed now, and we can compute the most important loop parameters. For the natural frequency and damping factor, we get from Eq. (3.11).

$$\omega_n = 251'000 \, \text{rad/s} (f_n = 40 \, \text{kHz})$$

$$\zeta = 0.5$$

From (6.10), the lock range becomes

$$\Delta\omega_L = 197'820 \, \text{rad} \, s(\Delta f_L = 31.5 \, \text{kHz})$$

and from (6.11), the lock time becomes

$$T_L = 25 \, \mu s$$

All block parameters have been determined now in the complex s domain. To get a digital Costas loop, we must convert the transfer function in the s domain into transfer functions in the z domain, using the z transform. As done in Sect. 3.6, a suitable sampling frequency f_{samp} must be chosen. As shown previously, f_{samp} must be greater than 4 times the center frequency of the Costas loop. A suitable choice would be $f_{samp} = 8 \, f_0 = 3.2 \, \text{MHz}$.

Next the transfer functions of the building block have to be converted into discrete transfer functions, i.e., $H(s) \rightarrow H(z)$. For best results, it is preferable to use the bilinear z transform [6]. Given an analog transfer function H(s), this can be converted into a discrete transfer function H(z) by replacing s by

$$s = \frac{2}{T} \frac{1 - z^{-1}}{1 + z^{-1}}$$

Now the bilinear z transform has the property that the analog frequency range from 0 to ∞ is compressed to the digital frequency range from 0 to $f_{samp}/2$. To avoid undesired "shrinking" of the corner frequency ω_C, this must be "prewarped" accordingly, i.e., we must set

$$\omega_{C,p} = \frac{2}{T} tg \frac{\omega_C T}{2}$$

where $\omega_{C,p}$ is the prewarped corner frequency. Now we can apply the bilinear z transform to the transfer functions of the loop filter [cf. Eq. (3.5)] and get

$$H_{LF}(z) = \frac{\left[1 + \frac{2}{\omega_{C,p} T}\right] + \left[1 - \frac{2}{\omega_{C,p} T}\right] z^{-1}}{\frac{2\tau_1}{T} - \frac{2\tau_1}{T} z^{-1}}$$

Because the VCO is a simple integrator, we can apply the discrete z transform of an integrator, i.e.,

$$H_{VCO}(z) = \frac{K_0 T}{1 - z^{-1}}$$

The digital Costas loop is ready now for implementation. A Simulink model will be presented in Sect. 6.6.

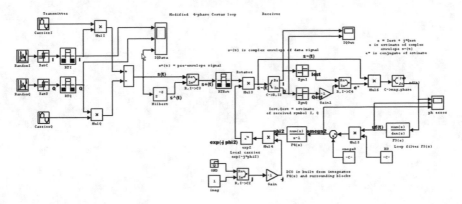

Fig. 6.8 Simulink model of the modified Costas loop for BPSK

Table 6.1 Comparison of predicted and simulated results for the pull-in range

Δf_0 (kHz)	$\Delta \omega_0$ (rad s^{-1})	T_P (theory) (µs)	T_P (simulation) (µs)
50	314,200	20	20
100	628,000	81	80
200	1,256,000	327	300

6.7   Simulating the Digital Costas Loop for BPSK

For the following simulation, the Simulink model **QPSK_Comp.mdl** is used.

Figure 6.8 shows the Simulink model of the Costas loop. Table 6.1 lists a number of results for the pull-in time T_P.

The predictions come very close to the results obtained from the simulation.

References

1. S.A. Tretter, *Communicatiions System Design Using DSP Algorithms*, (Springer Science + Business Media, 2008)
2. R.E. Best, N.V. Kuznetsov, G.A. Leonow, M.V. Yuldashev, R.V. Yuldashev, Simulation of analog Costas loop circuits. Int. J. Autom. Comput. 571–579 (Dec 2014)
3. R.E. Best, N.V. Kuznetsov, G.A. Leonov, M.V. Yuldashev, R.V. Yuldashev, A short survey on nonlinear models of the classic Costas loop: rigorous derivation and limitation of the classic analysis. American control conference, (Chicago, IL, USA, 1–3 July 2015), pp. 1296-1302
4. R.E. Best, N.V. Kuznetsov, G.A. Leonov, M.V. Yuldashev, R.V. Yuldashev, Tutorial on dynamic analysis of the Costas loop. Annu. Rev. Control 42, 27–49 (2016) (Elsevier)
5. H. Meyr, M. Moeneclaey, S.A. Fechtel, *Digital Communication Receivers*, 2nd edn. (Wiley & Sons, Inc, 1997)
6. E. Roland, *Best, Phase-locked Loops, Design, Simulation, and Applications*, 6th edn. (McGraw-Hill, New York, 2007)

Chapter 7
Costas Loop for m-ary Phase Shift Keying (mPSK)

7.1 Operating Principle

In the previous chapters, we discussed Costas loops for binary and quadrature phase shift keying. In BPSK, one bit is transmitted at a time, and in QPSK, two bits can be transmitted in one symbol. This technique can be extended to transmit more than 2 bits per symbol when the number of phase constellations is increased. This is called m-ary PSK [1, 2]. Here, the phase of the modulated signal can have one of the m states. In the following, we consider an example for m = 8, i.e., the phase of the modulated signal can be either 0, π/4, 2 π/4, 3 π/4 ... 7 π/4. The block diagram of a Costas loop for m-ary PSK is shown in Fig. 7.1.

In m-ary PSK, the transmitter generates a signal

$$s(t) = \cos(\omega_1 t + \frac{2\pi}{m} k) \quad with\, k = 0 \ldots m - 1$$

where ω_1 is the radian frequency of the carrier, and k is an integer in the range 0 ... m − 1. When m = 8 is chosen, the modulated signal can have 8 different phase constellations. The circuit shown in Fig. 7.1 is a modified Costas loop, i.e., it works with pre-envelope signals. We are considering this type of Costas loop because it is easier to implement and has better dynamic performance than conventional Costas loops. The pre-envelope signal then becomes

$$s^+(t) = s(t) + jH[s(t)] = \cos(\omega_1 t + \frac{2\pi}{m} k) + j \sin(\omega_1 t + \frac{2\pi}{m} k)$$
$$= \exp(j[\omega_1 t + \frac{2\pi}{m} k])$$

Electronic supplementary material The online version of this chapter (doi:10.1007/978-3-319-72008-1_7) contains supplementary material, which is available to authorized users.

R. Best, *Costas Loops*, https://doi.org/10.1007/978-3-319-72008-1_7

Fig. 7.1 Block diagram of
modified Costas loop for
mPSK

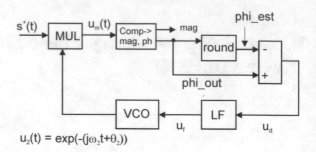

$$u_2(t) = \exp(-(j\omega_2 t + \theta_2))$$

where H(..) stands for Hilbert transform. The local oscillator (VCO) must then
generate a complex signal of the form

$$u_2(t) = \exp(-[\omega_2 t + \theta_2])$$

where θ_2 is the initial phase. This signal is multiplied in the block MUL with the
input signal $s^+(t)$; hence, the output signal of the multiplier is

$$u_m(t) = \exp(j[\omega_1 - \omega_2]t + j\frac{2\pi}{m}k - j\theta_2)$$

When the loop is locked in both frequency and phase, $\omega_1 = \omega_2$, and $\theta_2 \approx 0$; hence,
we have

$$u_m(t) \approx \exp\left(j\frac{2\pi}{m}k\right)$$

and the phase of $u_m(t)$ is an integer multiple of $2\pi/m$. The phase of $u_m(t)$ is labeled
as phi_out in Fig. 7.1. When a phase error θ_e exists, it is computed from

$$\theta_e = phi\_out - phi\_est$$

where phi_est is an exact multiple of $2\pi/m$. phi_est is obtained by the block entitled
"round" in Fig. 7.1. phi_est is an estimate of the actual phase and is the value that is
closest to the actual phase phi_out. This phase error is actually the output signal u_d
of a phase detector and is applied to the input of the loop filter LF. The output signal
u_f of the loop filter controls the frequency of the VCO. As the signal ud is identical
with the phase error, this phase detector has phase detector gain $K_d = 1$.

7.2 Transfer Function of the modified Costas Loop
for mPSK

We now can compute the transfer function H(s) of the modified Costas loop for
mPSK. A linear model of the loop is shown in Fig. 7.2.

Fig. 7.2 Linear model of the
modified Costas loop for
mPSK

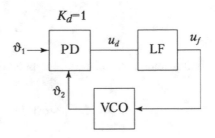

The transfer functions of the loop filter and VCO have been defined in Eqs. (3.5) and (3.6). Note that with this type of Costas loop there is no additional lowpass filter because the multiplication of the two complex carriers [cf. Eq. (6.6)] does not create the unwanted double-frequency component as found with the conventional Costas loops. From the model of Fig. 7.2, the open loop transfer function is determined to be

$$G_{OL}(s) = K_d \frac{K_0}{s} \frac{1 + s\tau_2}{s\tau_1}$$

Figure 7.3 shows a Bode plot of the magnitude of G_{OL}. The plot is characterized by the corner frequency ω_C which is defined by $\omega_C = 1/\tau_2$, and gain parameters K_d and K_0. At lower frequencies, the magnitude rolls off with a slope of -40 dB/decade. At frequency ω_C, the zero of the loop filter causes the magnitude to change its slope to -20 dB/decade. To get a stable system, the magnitude curve should cut the 0 dB line with a slope that is markedly less than -40 dB/decade. Setting the parameters such that the gain is just 0 dB at frequency ω_C provides a phase margin of 45 degrees which assures stability [3]. From the open loop transfer function, we now can calculate the closed loop transfer function defined by

Fig. 7.3 Bode plot of
magnitude of open loop gain
$G_{OL}(\omega)$

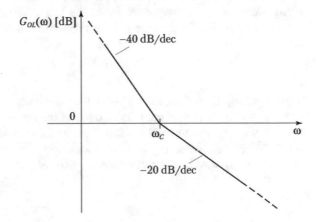

$$G_{CL}(s) = \frac{\Theta_2(s)}{\Theta_1(s)}$$

After some mathematical manipulations, we get

$$G_{CL}(s) = \frac{K_0 K_d \frac{1+s\tau_2}{s\tau_1}}{s^2 + s\frac{K_0 K_d \tau_2}{\tau_1} + \frac{K_0 K_d}{\tau_1}}$$

It is customary to represent this transfer function in normalized form, i.e.,

$$G_{CS}(s) = \frac{2s\zeta\omega_n + \omega_n^2}{s^2 + 2s\zeta\omega_n + \omega_n^2}$$

with the substitutions

$$\omega_n = \sqrt{\frac{K_0 K_d}{\tau_1}} \quad , \quad \zeta = \frac{\omega_n \tau_2}{2} \tag{7.1}$$

where ω_n is called *natural frequency*, and ζ is called *damping factor*. The linear model enables us to derive simple expressions for lock range $\Delta\omega_L$ and lock time T_L.

7.3 Lock Range $\Delta\omega_L$ and Lock Time T_L

For the following analysis, we assume that the loop is initially out of lock. The frequency of the reference signal (Fig. 7.1) is ω_1, and the frequency of the VCO is ω_2. The output signal of multiplier MUL is then a phasor rotating with angular velocity $\Delta\omega = \omega_1 - \omega_2$. Consequently the phase output of block "Complex $\rightarrow$ mag, phase" is a sawtooth signal having amplitude $(\pi/8)\, K_d$ and fundamental frequency $8\,\Delta\omega$, as shown in the upper trace of Fig. 7.4. As $8\,\Delta\omega$ is usually much higher than the corner frequency ω_C of the loop filter, the transfer function of the loop filter at higher frequencies can be approximated again by

$$H_{LF}(\omega) \approx \frac{\tau_2}{\tau_1} = K_H$$

The output signal u_f of the loop filter is a sawtooth signal as well and has amplitude $(\pi/8)\, K_d K_H$, as shown in the middle trace of the figure. This signal modulates the frequency ω_2 generated by the VCO. The modulation amplitude is given by $(\pi/8)$ $K_d K_H K_0$, cf. bottom trace. The Costas loop spontaneously acquires lock when the peak of the ω_2 waveform touches the ω_1 line; hence, we have

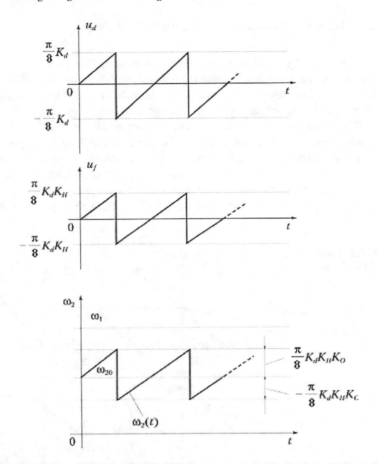

Fig. 7.4 Signals u_d, u_f, and ω_2 during the lock process

$$\Delta\omega_L = \frac{\pi}{8} K_d K_0 K_H = \frac{\pi}{8} K_d K_0 \frac{\tau_2}{\tau_1}$$

Making use of the substitutions Eq. (7.1) this can be rewritten as

$$\Delta\omega_L = \frac{\pi}{4} \zeta \omega_n \qquad (7.2)$$

As the lock process is a damped oscillation having frequency ω_n, the lock time can be approximated by one cycle of this oscillation, i.e.,

$$T_L \approx \frac{2\pi}{\omega_n} \qquad (7.3)$$

7.4   Nonlinear Model for the Unlocked State

To derive the model for the unlocked state, we assume that the loop is not yet
locked, and that the difference between reference frequency ω_1 and VCO output
frequency ω_2 is $\Delta\omega = \omega_1 - \omega_2$. As shown in Sect. 6.2 (cf. also Fig. 6.5), u_d is a
sawtooth signal having frequency 8 $\Delta\omega$ (cf. upper trace in Fig. 7.5).

As will be explained in short, this signal is asymmetrical, i.e., the duration of the
positive which has wave T_1 is not identical with the duration T_2 of the negative. The
middle trace shows the output signal of the loop filter, and the lower trace shows the
modulation of the VCO output frequency ω_2. From this waveform, it is seen that
during T_1 the average frequency difference $\Delta\omega$ becomes smaller, but during interval
T_2 it becomes larger. Consequently the duration of T_1 is longer than the duration of

Fig. 7.5 Pull-in process of
the modified Costas loop for
mPSK

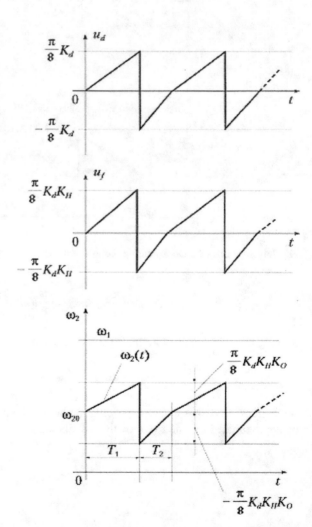

T_2, and the average of signal u_d is non-zero and positive. Using the same mathematical procedure as in Sects. 3.2 and 4.3, the average $\overline{u_d}$ can be computed from

$$\overline{u_d} = \frac{\pi^2 K_d^2 K_0 K_H}{128\,\Delta\omega} \tag{7.4}$$

Because this type of Costas loop does not require an additional lowpass filter, the u_d signal is not shifted in-phase, and therefore, there is no cos term in Eq. (6.12). This implies that there is no polarity reversal in the function $\overline{u_d}(\Delta\omega)$, hence the pull-in range becomes theoretically infinite. Of course in a real circuit, the pull-in range will be limited by the frequency range the VCO is capable to generate. When the center frequency f_0 of the loop is 10 MHz, for example, and when the VCO can create frequencies in the range from 0 ... 20 MHz, then the maximum pull-in range Δf_P is 10 MHz, i.e., $\Delta\omega_P = 6.28\ 10^6$ rad/s.

Figure 7.6 shows the nonlinear model used to compute the pull-in process. The block generating signal $\overline{u_d}$ is labeled "Phase/Frequency Detector" [cf. Eq. (5.10)] because in the unlocked state $\overline{u_d}$ is a function of the frequency error $\Delta\omega$.

Using the same mathematical procedure as in Sect. 3.3, the transfer function of the loop filter can be approximated as

$$H_{LF}(s) \approx \frac{1}{s\,\tau_1} \tag{7.5}$$

In time domain, we can therefore write

$$\overline{u_f}(t) = \frac{1}{\tau_1}\int_0^t \overline{u_d}(\tau)\,d\tau \tag{7.6}$$

For the transfer function of the VCO, we get

$$\Delta\omega = \Delta\omega_0 - K_0\overline{u_f} \tag{7.7}$$

where $\Delta\omega_0$ is the initial frequency difference given by

$$\Delta\omega_0 = \omega_1 - \omega_{20}$$

Fig. 7.6 Nonlinear model of
the modified Costas loop

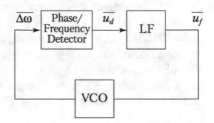

Equations (7.4) … (7.7) enable us to compute the pull-in process, i.e., the instantaneous frequency difference $\Delta\omega$ as a function of time. This will be demonstrated in the next section.

7.5 Pull-in Range and Pull-in Time of the Modified Costas Loop for QPSK

As seen in the last section, the pull-in range of this type of Costas loop can be arbitrarily large. Using Eqs. (6.13), (6.15), and (6.16), we can derive an equation for the pull-in time T_P:

$$T_P \approx \frac{32}{\pi^2} \frac{\Delta\omega_0^2}{\zeta\,\omega_n^3} \tag{7.8}$$

7.6 Design Procedure for Costas Loop for mPSK

In the following, a design procedure for a digitally modified Costas loop is presented. The design is based on the method we already used in Sect. 5.5. It is assumed that two binary signals (I and Q) are modulated onto a quadrature carrier (cosine and sine carrier). The carrier frequency is set to 400 kHz, i.e., the Costas loop will operate at a center frequency $\omega_0 = 2\,\pi\,400'000 = 2'512'000$ rad s^{-1}. The symbol rate is assumed to be $f_S = 100'000$ symbols/s. Now the parameters of the loop (such as time constants τ_1 and τ_2, corner frequency ω_C, and gain parameters such as K_0, K_d) must be determined. (Note that these parameters have been defined in Eqs. (3.4–3.6), and (3.13).

It has been shown in Sect. 6.1 that for this type of Costas loop, $K_d = 1$. It has proven advantageous to determine the remaining parameters by using the open loop transfer function $G_{OL}(s)$ of the loop [3], which is given here by

$$G_{OL}(s) = K_d \frac{K_0}{s} \frac{1 + s\tau_2}{s\tau_1}$$

The magnitude of $G_{OL}(\omega)$ has been shown in Fig. 7.3. As already explained in Sect. 3.5, the magnitude curve crosses the 0 dB line at the transit frequency ω_T. As in the case of the Costas loop for BPSK, we again set $\omega_T = 0.1\,\omega_0$, i.e., $\omega_T = 251'200$ rad s^{-1}. Furthermore, we set corner frequency $\omega_C = \omega_T$. When doing so, the slope of the asymptotic magnitude curve changes from -40 dB/decade to -20 dB/decade at $\omega = \omega_C$. Under this condition, the phase of $G_{OL}(\omega)$ is $-135°$ at ω_C. Consequently, the phase margin of the loop becomes 45° which provides sufficient stability. According to Eq. (3.5), τ_2 becomes 4 µs. Lastly, the remaining parameters τ_1 and K_0 must be chosen. They have to be specified such that the open loop gain becomes 1 at frequency $\omega = \omega_C$. According to Eq. (3.5), we can set

$$G_{OL}(\omega_C) = 1 \approx \frac{K_0 \, K_d}{\omega_c^2 \, \tau_1}$$

Because 2 parameters are still undetermined, one of those can be chosen arbitrarily, hence we set $\tau_1 = 20$ µs. We then get $K_0 = 1'262'000 \ \text{s}^{-1}$.

The design of the Costas loop is completed now, and we can compute the most important loop parameters. For the natural frequency and damping factor we get from Eq. (3.11),

$$\omega_n = 251'000 \, \text{rad/s} \ (f_n = 40 \, \text{kHz})$$
$$\zeta = 0.5$$

From (7.3), the lock range becomes

$$\Delta\omega_L = 98'500 \, \text{rad s}^{-1}(\Delta f_L = 15.7 \text{kHz})$$

and from (6.11) the lock time becomes

$$T_L = 25 \ \mu s$$

All block parameters have been determined now in the complex s domain. To get a digital Costas loop, we must convert the transfer function in the s domain into transfer functions in the z domain, using the z transform. As done in Sect. 3.6, a suitable sampling frequency f_{samp} must be chosen. As shown previously, f_{samp} must be greater than 4 times the center frequency of the Costas loop. A suitable choice would be $f_{samp} = 8$ and $f_0 = 3.2$ MHz.

Next, the transfer functions of the building block have to be converted into discrete transfer functions, i.e., $H(s) \rightarrow H(z)$. For best results, it is preferable to use the bilinear z transform [3]. Given an analog transfer function H(s), this can be converted into a discrete transfer function H(z) by replacing s by

$$s = \frac{2}{T}\frac{1 - z^{-1}}{1 + z^{-1}}$$

Now, the bilinear z transform has the property that the analog frequency range from $0 \dots \infty$ is compressed to the digital frequency range from $0 \dots f_{samp}/2$. To avoid undesired "shrinking" of the corner frequency ω_C, this must be "prewarped" accordingly, i.e., we must set

$$\omega_{C,p} = \frac{2}{T} tg \frac{\omega_C \, T}{2}$$

where $\omega_{C,p}$ is the prewarped corner frequency. Now, we can apply the bilinear z transform to the transfer functions of the loop filter [cf. Eq. (3.5)] and get

$$H_{LF}(z) = \frac{\left[1 + \frac{2}{\omega_{C,p}T}\right] + \left[1 - \frac{2}{\omega_{C,p}T}\right]z^{-1}}{\frac{2\tau_1}{T} - \frac{2\tau_1}{T}z^{-1}}$$

As the VCO is a simple integrator, we can apply the discrete z transform of an integrator, i.e.,

$$H_{VCO}(z) = \frac{K_0 T}{1 - z^{-1}}$$

The phase detector used in this Costas loop is somewhat special, because the phase error is computed on the base of estimates of the actual phase constellation. Details about phase detector design are presented in Sect. 7.7, where we will perform simulations with that type of Costas loop.

7.7   Simulink Model for Costas Loop for mPSK

A Simulink model for a Costas loop for mPSK is shown in Fig. 7.7 (mPSK_Comp. mdl). The model consists of a transmitter and a receiver circuit. In the following, various parts of the model are described.

(a) Transmitter

In the transmitter section of the model, an 8-ary random signal k(n) is created, which is in the range $-4 \dots 3$. (n = sample index). Then a complex phasor exp(j k (n) pi/4) is computed. Data signal I is then given by the real part, and data signal Q is given by the imaginary part of that phasor. This phasor has a magnitude of 1 and can have 1 of 8 possible phases, i.e., 0, pi/4, pi/2, 3 pi/4, etc.

Furthermore, a complex carrier exp(j omegaC t + theta1) = exp(j phi1) is generated. (In this simulation, theta1 is set 0.) Multiplying the phasor with the complex carrier yields the pre-envelope signal s + (t)

$$s + (t) = (I + jQ)^* \exp(j\, omegaC^* t)$$

The transmitter output signal s(t) is by definition the real part of s + (t)

$$s(t) = Re[s + (t)] = I \cos(omegaC\, t) - Q \sin(omegaC\, t)$$

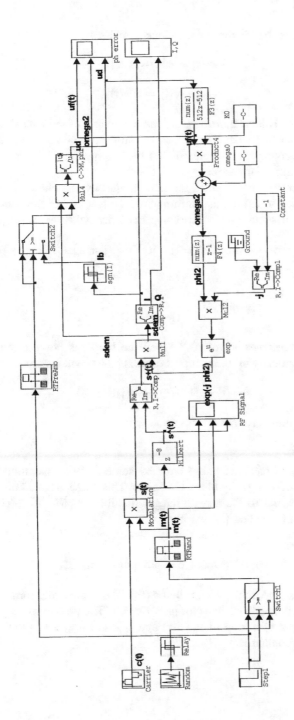

Fig. 7.7 Block diagram of Simulink model mPSK_Comp.mdl

(b) **Receiver**

The receiver first builds the pre-envelope signal s + (t)

$$s + (t) = s(t) + j s \wedge (t)$$

$$\text{with } s \wedge (t) = H[s(t)]$$

where H[.] denotes Hilbert transform. As the bandwidth of the data signal is much less than the carrier frequency, the Hilbert transform can be computed by a simple delay, i.e., by delaying the samples of s(t) by one quarter of a carrier period (8 samples by default in our case).

As the receiver does not "know" the exact frequency and phase of the carrier, a Costas loop is used to extract that information. To reconstruct the data signal, the pre-envelope signal is multiplied with a complex carrier of the form exp(−j omegaC t):

$$s \sim (t) = s + (t) \exp(-j \, \text{omegaC} \, t) = (I + jQ)^* \exp(j \, \text{omegaC} \, t)^* \exp(-j \, \text{omegaC} \, t)$$

This yields

$$s \sim (t) = I + j Q$$

which is called "complex envelope." When the loop has not yet acquired lock, s ~ (t) is not identical with the complex envelope but contains a phase error term:

$$s \sim (t) = (I + j Q)^* \exp(j \, \text{theta\_e})$$

where theta_e is the phase error.

(c) **Phase detector**

We now have to extract that phase error by some suitable operations. First, we extract the phase of s ~ (t) by block C->M,phi. This phase is called phi_out. Next, we round that phi_out to the nearest integer multiple of pi/4 (45°). This yields a phase estimate and is called phi_est here.

Now, the phase error theta_e is simply the difference

$$ud(t) = \text{theta\_e} = \text{phi\_out} - \text{phi\_est}$$

The blocks Mul3, C->M,phi, *4/pi, round, *(−pi/4) end the adder form; therefore, a phase detector that has phase detector gain Kd = 1. The phase error signal ud(t) is applied to the input of a loop filter F3(z) that is realized as a PI filter. Its transfer function in the s domain is

$$F3(s) = \frac{1 + s\,tau2}{s\,tau1}$$

This transfer function is converted into F3(z) by the bilinear z transform (cf. the comments in initialization file InitFcnmPSK_Comp.m).

The output of the loop filter is applied to the input of the DCO built from F4(z) and surrounding blocks. In the s domain, the transfer function of the DCO is given by

$$F4(s) = \frac{K0}{s}$$

with K0 = DCO gain. This transfer function is also converted into F4(z) by using the bilinear z transform (cf. the comments in initialization file InitFctMPSK8D.m).

The radian frequency of the DCO is given by

$$omega2 = omega0 + K0\,uf$$

where uf is the output signal of the loop filter. omega0 is the center radian frequency of the Costas loop. Ideally, omega0 should be equal with omegaC. It is allowed, however, that omega0 deviates from omegaC, because the Costas loop can track the frequency error.

(d) **Parameters of the model**

A number of parameters can be specified by the operator:

- fC carrier frequency of the transmitter (default = 400'000 Hz).
- fR symbol rate (default = 100'000 symb/s).
- OS oversampling factor factor used in the transmitter section. The sampling frequency of the model is defined as the product OS * fC (default = 32).
- nCycles number of symbols used in the simulation (default = 20).
- delta_f frequency error of receiver carrier frequency. To simulate a frequency offset, the initial frequency of the Digital-controlled oscillator (DCO) is set to fC − delta_f.
- D decimation factor (default = 8). This allows to sample the blocks within the receiver with a lower sampling frequency.

(e) **Instructions for model operation**

To perform simulations, proceed as follows:

- load the model mPSK phase shift keying (mPSK)_Comp by double-clicking the file QPSK Phase Shift Keying (QPSK)_Comp.mdl in the Current Folder of MATLABMatlab.

– this displays the model and a figure window containing 6 edit controls for parameter specification. When the model runs the first time, default parameters are set (cf. Sect. 7.2). You now can alter these parameters. When done, click the "Done" button. When an invalid number format has been entered in one of the edit controls, an error message is issued, telling you to correct this entry and to hit the "Done" button again. Hitting this button saves the actual parameters to a parameter file params_mQPSK_Comp.mat. When the model is started next, the parameters saved in these files are loaded.

There is an option to load the initial default parameters: hit the "Set init defaults" button.

This can be useful whenever you specify parameters that do not give useful results.

– after hitting the "Done" button, go to the model window and start the simulation (menu item Simulation/Start).
– look at the results on the scopes phi_in and ph error. phi_in represents the phase of the data phasor I + j Q. k(n) can have integer values in the range −4... 3, and the corresponding phase is k(n) * pi/4. The reconstructed phase is seen in trace k^(n)wrap in scope phi error. The Wrap block is used to get the same phase readout at the input and at the output of the model. As the round block can also yield a result of +4, this is wrapped to −4 in order to have the same scales for input and output phase. From the theta_e, uf, and omega2 signals, you can see how fast the lock acquires lock.

(f) **Comment on the pull-in range of the Costas loop**

In contrast to the conventional Costas loop (as used in model QPSK_Real), the modified Costas loop does not require an additional lowpass filter to remove the unwanted double-frequency components. Hence, there is no additional phase shift in the Costas loop, and the pull-in range can be arbitrarily high. When selecting a higher delta_f value, it will be necessary to increase the duration of the simulation (increase nCycles), because the pull-in time TP will become larger.

When very large frequency errors are simulated, you will note that the loop is no longer able to acquire lock when larger decimation factors are used. This stems from the fact that the output signal of block Mul3 will contain very high frequency components, i.e., waveforms whose frequency is up to 8 times the frequency error. To process those high-frequency signals, a large sampling rate must be selected; otherwise, we will be confronted with aliasing effects.

(g) **Comment on the "phase ambiguity" of the Costas loop**

When performing simulations with different values of frequency error delta_f, you will recognize in some situations that the phase of the output signal can be offset from the phase of the input signal.

This occurs because the Costas loop for mQPSK can lock with 8 possible phase differences between transmitter and receiver carriers, i.e., with a phase difference of 0, 45, 90, 135, etc., degrees.

To avoid this ambiguity, additional measures have to be taken. A common method is to use a given preamble at each start of a data transmission, e.g., a sequence of symbols having all the same phase.

Because the receiver knows what symbols are sent in the preamble, it will replace the demodulated I signal with these symbols during the interval where the preamble is transmitted. This method is demonstrated in an other model (BPSK_Real_
PreAmb).

The same procedure could be applied in this model, i.e., we would have to create a predefined preamble for both I and Q signals, e.g., a sequence of all 1's.

References

1. M. Floyd, *Gardner, Phase-lock Techniques*, 2d edn. (Wiley, New York, 1979)
2. B. Sklar, *Digital Communications* (Prentice Hall, Fundamentals and Applications, 1988)
3. E.B. Roland, *Phase-locked Loops, Design, Simulation, and Applications*, 6th edn. (McGraw-Hill, New York, 2007)

Chapter 8
Costas Loop for BPSK Using Phasor Rotator Circuit

8.1 Operating Principle

In this chapter, we consider another variant of the Costas loop that does not use a voltage-controlled oscillator (VCO) or digital-controlled oscillator (DCO), but acquires the locked state by rotating a phasor [1]. The block diagram of this Costas loop is shown in Fig. 8.1.

The blocks at the left of the diagram are identical with those of the Costas loop considered in Chap. 3, cf. Fig. 3.1a. The input signal is given by

$$s(t) = m \, \sin(\omega_1 t + \theta_1)$$

In place of the VCO in Fig. 3.1a, an ordinary oscillator (Osc) is used that generates a fixed frequency ω_2. This oscillator has a sine and a cosine output. The output signals of the multipliers in the I branch and Q branch are again lowpass filtered to remove the high-frequency component at about twice the carrier frequency. The output signal of the lowpass filter in the I branch is therefore

$$I = m \, \cos([\omega_1 - \omega_2] \, t + \theta_1 - \theta_2),$$

and the output signal of the lowpass filter in the Q branch becomes

$$Q = \sin([\omega_1 - \omega_2] \, t + \theta_1 - \theta_2)$$

Electronic supplementary material The online version of this chapter (doi:10.1007/978-3-319-72008-1_8) contains supplementary material, which is available to authorized users.

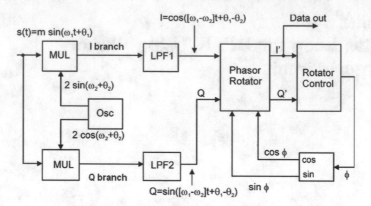

Fig. 8.1 Costas loop for BPSK using phasor rotator

We now define a phasor P by

$$P = I + jQ = m \, \cos([\omega_1 - \omega_2] \, t + \theta_1 - \theta_2) + j \, m \, \sin([\omega_1 - \omega_2] \, t + \theta_1 - \theta_2)$$
$$= m \, \exp(j[\Delta\omega \, t + \theta_e])$$

$$(8.1)$$

with $\Delta\omega = \omega_1 - \omega_2$ and $\theta_e = \theta_1 - \theta_2$, i.e., signal I is considered to be the real part and signal Q the imaginary part of phasor P. Assume for the moment that m can take the values = 1 or −1. According to Eq. (8.1), P rotates with a radian frequency $\Delta\omega$, i.e., it executes $\Delta\omega/2 \, \pi$ revolutions per second. But our aim is to get the loop locked, which means that the phasor P should settle at a position P = 1 + j · 0 when m = 1 or at a position P = −1 + j · 0 when m = −1. To reach that goal, we must therefore rotate phasor P in opposite direction. When $\Delta\omega$ is positive, e.g., P rotates in positive (counterclockwise) direction. In this case, the phasor rotator must rotate P in the negative direction, i.e., clockwise. The operating principle of the phasor rotator is explained by Fig. 8.2.

The rotator can be considered as a rotating switch. It can rotate the phase of phasor P by integer multiples of a phase step $\Delta\phi$. $\Delta\phi$ has been chosen $2 \, \pi/16$ in this example, which corresponds to an angle of 22.5°. The choice of the phase step $\Delta\phi$ is crucial for the operation of the loop. We will consider the criteria for selecting the appropriate value for $\Delta\phi$ later in this section.

As shown in Fig. 8.2, the phasor P can be rotated by 0°, 22.5°, 45°, 67.5° … 337.5°.

The position of the rotating switch is controlled by a circuit labeled *Rotator Control*, cf. Fig. 8.3.

To determine the required direction of ratation, it is checked first in which of the four quadrants the rotated phasor P′ is currently positioned, cf. Fig. 8.5. When P′ is in quadrant Q4, it must be rotated in positive (counterclockwise) direction in order to get settled at a position near P′ = 1 + j · 0. When P′ is currently in quadrant Q1,

Fig. 8.2 Phasor rotator operation

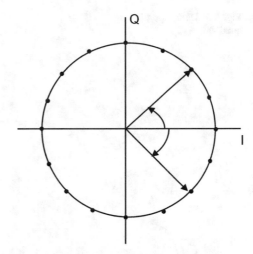

Fig. 8.3 Operation of the Rotator Control circuit

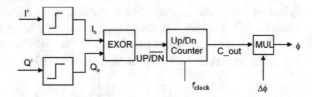

Fig. 8.4 Operation of the phasor rotator circuit

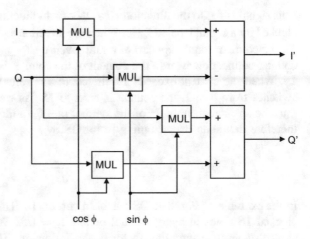

however, it must be rotated in negative (clockwise) direction to get settled at $P' = 1 + j \cdot 0$. When P' is in quadrant $Q2$, it must be rotated positively in order to get settled at $P' = P' = -1 + j \cdot 0$. When P' is in quadrant $Q3$, it must be rotated negatively to reach a position near $P' = -1 + j \cdot 0$. It turns out that the required direction is positive whenever I' and Q' have opposite sign, and negative, whenever I' and Q'

Fig. 8.5 Definition of
quadrants

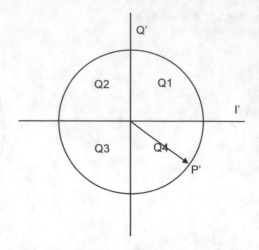

have the same sign. Therefore, two binary signals I_b and Q_b are created, cf. Fig. 8.3.
I_b is a logical 1 when the sign of I' is positive and a logical 0 when I' is negative. The
same holds for binary signal Q_b: Q_b is a logical 1 when Q' is positive. The rotation
direction is now positive when I' and Q' have different sign. Hence, we derive a
logical signal labeled $UP/\overline{DN}$ is built from the EXOR function of I_b and Q_b,

$$UP/\overline{DN} = EXOR(I_b, Q_b) = \overline{I_b} \cdot Q_b + I_b \cdot \overline{Q_b}$$

This signal is fed to the direction $UP/\overline{DN}$ of a bidirectional counter. This counter is
clocked by a signal labeled f_{clock}. When the direction signal $UP/\overline{DN}$ is high (logical
1), the counter counts upward at a rate given by f_{clock}. When $UP/\overline{DN}$ is low, the
counter counts downward. The content of the counter is restricted to the range 0...
15. When the content exceeds 15, the counter is reset to 0, and when the content
switches from 0 to 1, the content is reset to 15. As we will recognize soon, f_{clock}
must be an integer multiple of the symbol rate f_S (number of bits per second). We
therefore define an oversampling factor OS by

$$OS = \frac{f_{clock}}{f_S}$$

In the example of Fig. 8.1, OS has been chosen 16. This means that the counter is
clocked 16 times in every symbol period $T_S = 1/f_S$. We will discuss the optimal
choice of the oversampling factor in the following. The operation of the phasor
rotator becomes now evident: When phasor P' is in quadrant Q4, for example, the
counter in Fig. 8.3 counts upward. Every clock pulse applied to the counting input
of the counter increases the content C_out of the counter, and the rotation angle ϕ is
incremented by every clock pulse. This happens until the rotated phasor reaches the
point $P' = 1 + j \cdot 0$. The counter now counts in opposite direction, i.e., downward.

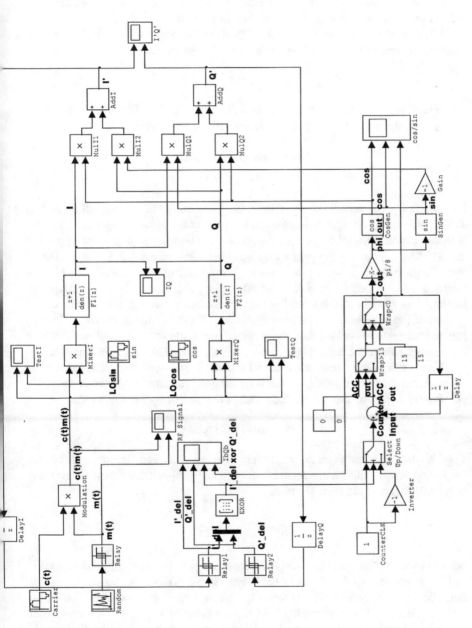

Fig. 8.6 Simulink model of Costas loop for BPSK using phasor rotator

This decreases the rotation angle ϕ. When the loop has acquired lock, the rotated phasor is positioned either near $P' = 1 + j \cdot 0$ when m = 1 or near $P' = 1 - j \cdot 0$ when m = -1. Because the rotated phasor P' can be incremented or decremented only in increments of $\Delta\phi$, P' settles within an error band of $\pm\Delta\phi$.

The content of the counter C_out determines the angle ϕ by which phasor P must be rotated. ϕ is obtained by multiplying C_out by ΔC

$$\phi = C\_out \cdot \Delta\phi$$

The operation of the phasor rotator is explained by Fig. 8.4. To rotate phasor P by an angle ϕ, the following trigonometric operations are required:

$$I' = I \cdot \cos\ \phi - Q \cdot \sin\phi$$
$$Q' = I \cdot \sin\phi + Q \cdot \cos\phi \tag{8.2}$$

Last but not least, we will analyze how the parameters OS (oversampling factor) and the phase step $\Delta\phi$ should be chosen. Let us start with phase step $\Delta\phi$. When $\Delta\phi$ is made very small, for instance 5°, the phasor can be very accurately positioned at the desired location, and the phase error band becomes very small, i.e., $\pm5°$. This is not a large benefit, however, because an error band of say $\pm22.5°$ can also be tolerated without any trouble. On the other side, $\Delta\phi$ cannot be made arbitrarily large. When we choose, e.g., $\Delta\phi = 90°$, P' executes large jumps whenever the content of the counter is increased or decreased. If $\Delta\phi$ is chosen too large, this can lead to instabilities, because the loop is not able to settle at the desired position. For this reason, it seams reasonable to choose a value of 22.5° or 11.25° for $\Delta\phi$.

The oversampling factor must also be chosen appropriately. Assume for the moment that the phasor rotator circuit is rotating permanently in the same direction, positive or negative. The rotator then rotates the phasor by an angle rad/s.

$$\theta = \Delta\phi \cdot OS \cdot f_S \tag{8.3}$$

This is the maximum phase angle per second that can be tracked by the loop. Apparently, this figure is identical with the pull-in range $\Delta\omega_P$. The pull-in range expressed in Hz then becomes

$$\Delta f_P = \frac{\Delta\phi}{2\pi} \cdot OS \cdot f_S \tag{8.4}$$

Eq. (8.4) shows that the pull-inrange is proportional to the oversampling factor OS and to the phase step $\Delta\phi$. When a large oversampling factor is chosen, the pull-in range becomes large as well, but when the phase step is made smaller, the pull-in range is decreased correspondingly. In the example of Fig. 8.1, we have chosen $\Delta\phi = 2\pi/16$ and OS = 16, hence the pull range (in Hz) equals the symbol rate f_S.

The designer of a Costas loop using phasor rotation would certainly be interested to know the pull-in time T_P. Because this system is highly nonlinear, it is very difficult to derive an equation for T_P. Simulations show that the loop acquires lock very rapidly, i.e., in only a few cycles of the clock signal f_{clock}, and that the loop locks within one only symbol period $T_S = 1/f_S$,

$$T_P < 1/f_S \qquad (8.5)$$

We recognize that the acquisition process of this type of Costas loop is considerably faster than that of the Costas loops that use a VCO or DCO and a loop filter (Fig. 3.1a).

8.2 Design Procedure for Costas Loop Using Phasor Rotator

The design of this type of Costas loop is very simple, because only a few parameters must be set:

- the phase step $\Delta\phi$,
- the oversampling factor OS,
- the 3 dB corner frequency ω_{3dB} of the lowpass filters LPF1 and LPF2 (cf. Fig. 8.1).

Assume that the carrier frequency is $f_C = 400$ kHz and the symbol rate is $f_S = 100'000$ bits/s. As discussed in Sect. 8.1, an appropriate value must be chosen for the phase step $\Delta\phi$. We have seen that $\Delta\phi = 2\pi/16$ (22.5°) is a good choice. The oversampling factor will be chosen using Eq. (8.4). Assuming that we want a pull-in range of 100 kHz, Eq. (8.4) yields an oversampling factor OS = 16. Last we determine the 3 dB corner frequency of the lowpass filters. The modulating signal m(t) is a square wave function. The largest fundamental frequency of that signal is $f_S/2$, when a bit sequence of the form 01010101... is transmitted. ω_{3dB} must therefore be chosen markedly larger than $f_S/2$, but also smaller than twice the radian carrier frequency ω_C, which is $2\pi \cdot 400'000 = 2'512'000$ rad/s. A good choice would be to set the 3 dB corner frequency twice the symbol rate, i.e., $\omega_{3dB} = 2 \cdot 2\pi \cdot f_S = 1'256'000$ rad/s.

The transfer function of the lowpass filters is given by

$$H_{LPF}(s) = \frac{1}{1 + s/\omega_{3dB}}$$

When the lowpass filters are realized as digital filters, we must convert the transfer function $H_{LPF}(s)$ into the discrete transfer function $H_{LPF}(z)$. For best results, it is preferable to use the bilinear z transform [2]. Given an analog transfer function H(s), this can be converted into a discrete transfer function H(z) by replacing s by

$$s = \frac{2}{T}\frac{1 - z^{-1}}{1 + z^{-1}} \qquad (8.6)$$

T is the sampling interval used for this digital filter, and $f_F = 1/T$ is the sampling frequency. Because the highest frequency component at the output of these filters is twice the carrier frequency $2\,f_C$, according to the Nyquist theorem the sampling frequency f_F must be chosen at least twice that frequency, i.e., $f_F > 1.6$ MHz. Now the bilinear z transform has the property that the analog frequency ranges from 0 to ∞ is compressed to the digital frequency ranging from $0 \ldots f_{samp}/2$. To avoid undesired "shrinking" of the corner frequencies (ω_C and ω_3), these must be "pre-warped" accordingly, i.e., we must set

$$\omega_{3dB,p} = \frac{2}{T} tg \frac{\omega_{3dB}\,T}{2} \tag{8.7}$$

where $\omega_{3dB,p}$ is the prewarped corner frequency. Now we can apply the bilinear z transform to the transfer functions of the lowpass filters and get

$$H_{LPF}(z) = \frac{\left[1 + \frac{2}{\omega_{3dB,p}\,T}\right] + \left[1 - \frac{2}{\omega_{3dB,p}\,T}\right] z^{-1}}{1 + z^{-1}} \tag{8.8}$$

A Costas loop using these parameters will be presented in Sect. 8.3.

8.3   Simulating the Costas Loop for BPSK Using Phasor Rotator

The Simulink model BPSK5.mdl shown in Fig. 8.6 represents a BPSK system built from a transmitter and a receiver. The model uses the design parameters used in the design example of Sect. 8.2. The transmitter section is shown on left top of the block diagram. A random generator creates a random binary signal m(t), and this signal is modulated to a carrier c(t). At the right, the receiver section is shown. The input signal of the receiver is multiplied by a sine wave (I branch) and a cosine wave (Q branch). The frequency of these signals is fixed. The output signals of the multipliers are lowpass filtered. The filtered signal represents the phasor $P = I + j\,Q$. Phasor P is applied to the phasor rotator, which is built from the four multipliers and two adders shown at the right. Signal I is converted by Relay1 (which is actually a comparator) to a logical signal labeled I'_del, representing the sign I = of I' and signal Q' is converted by Relay2 to a logical signal labeled Q'_del, representing the sign of Q'. These two logical signals are applied to an EXOR gate, and the output of the EXOR is the

logical $UP/\overline{DN}$ signal that determines the counting direction of the Up/Down counter, built from an adder labeled Counter ACC.

To see how this type of Costas loop actually works, it is instructive to have a look on some waveforms, cf. Fig. 8.7. In this simulation, the frequency error is set 50 kHz. Figure 8.7 shows the data signal m(t) and the BPSK signal c(t) m(t), where

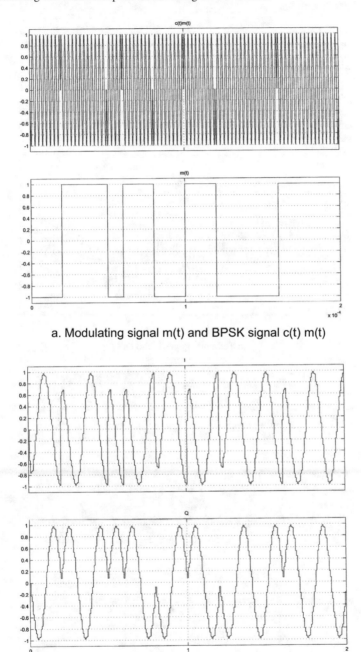

a. Modulating signal m(t) and BPSK signal c(t) m(t)

b. Output signals I and Q of the lowpass filters LPF1 and LPF2

Fig. 8.7   Signals of the Simulink model of Fig. 8.6

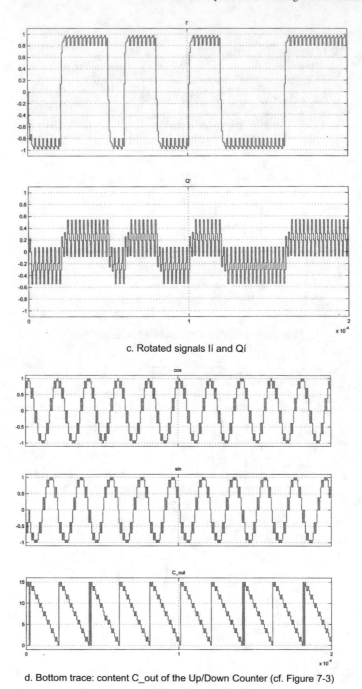

c. Rotated signals Ií and Qí

d. Bottom trace: content C_out of the Up/Down Counter (cf. Figure 7-3)

Fig. 8.7 (continued)

c(t) is the carrier. Figure 8.7b shows the output signals I and Q of the lowpass filters LPF1 and LPF2. The sum $I + j \cdot Q$ builds a phasor P rotating with a frequency of 50 kHz. The phasor rotation is now canceled by the phasor rotator. The rotated signals I′ and Q′ are shown in Fig. 8.7c. We recognize that I′ signal is perfectly aligned with the modulating signal m(t). Finally in Fig. 8.7d, the content of the Up/Down Counter C_out is shown, cf. bottom trace. This signal looks like a sawtooth signal. It periodically ramps from 15 down to 0 and restarts at 15 again, and the range of 0…15 corresponds to a phase shift in the range from 337.5° to 0° (Fig. 8.6).

8.4 Modified Costas Loop for BPSK Using Phasor Rotator

In the preceding section, we considered Costas loop for BPSK using phasor rotators that worked with a real input signal s(t). This type of Costas can be realized, however, also as modified Costas loop, i.e., a Costas loop working with pre-envelope signals.

The block diagram of such a loop is shown in Fig. 8.8.

The input signal is given by

$$s(t) = m \cos(\omega_1 t + \theta_1)$$

with m = modulating signal, ω_1 = reference frequency, and θ_1 = initial phase. m can have two equal and opposite values, either +1 and −1, or +c and −c, where c can be an arbitrary constant. The input signal is first converted into a pre-envelope signal, as explained in Sect. 1.1. The output signal of the Hilbert transformer is

$$\hat{s}(t) = H[m \cos(\omega_1 t + \theta_1)] = m \sin(\omega_1 t + \theta_1)$$

H(…) stands for Hilbert transform. (Note that because the largest frequency of the spectrum of the data signal m is much lower than the carrier frequency ω_1, the

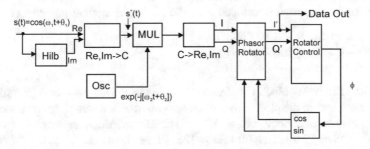

Fig. 8.8 Block diagram of modified Costas loop for BPSK using phasor rotator

Hilbert transform of the product $H[m \cos(\omega_1 t + \theta_1)]$ equals $m H[\cos(\omega_1 t + \theta_1)]$ [3].) The pre-envelope signal is obtained now from

$$s^+(t) = s(t) + j\hat{s}(t) = m \exp(j[\omega_1 t + \theta_1])$$

The ocsillator generates a complex signal

$$u_2(t) = \exp(-j[\omega_2 t + \theta_2])$$

The signals $s^+(t)$ and $u_2(t)$ are multiplied in block MUL. The output signal of MUL is

$$m \exp(j[(\omega_1 - \omega_2)t + \theta_1 - \theta_2]) = m \exp(-j[\Delta\omega t + \theta_e])$$

with $\Delta\omega = \omega_1 - \omega_2$ and $\theta_e = \theta_1 - \theta_2$. This is a phasor rotating with a frequency of $\Delta\omega$. The output signal of the multiplier is split by the block labeled C->Re,Im (complex to real, imaginary) into the two components I and Q. The remaining blocks of the loop are identical with those of the formerly discussed circuit of Fig. 8.1. The phasor $P = I + j \cdot Q$ is applied to the phasor rotator, and the position of the rotating switch is again controlled by block Rotator Control.

8.5 Simulating the Modified Costas Loop for BPSK Using Phasor Rotator

A data transmission system using this modified Costas loop has been realized by Simulink model BPSK6.mdl, cf. Fig. 8.9. Most blocks are identical with those of the model in Fig. 8.6. The transmitter generates directly the pre-envelope signal $s^+(t)$. The local oscillator is built from a sine and a cosine generator that generates a fixed frequency signal. These two signals are combined by block Re,Im ->C (Real, Imag to Complex) to form the complex signal $\exp(-j[\Delta\omega t + \theta_e])$. This signal is multiplied by the block labeled product. All remaining blocks are identical with those of the previous model (Fig. 8.6).

This model has been developed with the following default parameters:

- phase step $\Delta\phi = 2\pi/16$ (22.5°),
- symbol rate $f_S = 100'000$ bits/s,
- carrier frequency $f_C = 400$ kHz.

The blocks following block C->Re,Im (Complex to Real, Imaginary) (including the up/down counter in the phasor control circuit) operate at a sampling frequency of 32 times to carrier frequency, i.e., at 12.8 MHz. Hence, the oversampling factor OS becomes OS = 12.8 MHz/400 kHz = 128. This leads to a pull-in frequency $\Delta f_P = 800$ kHz. When running the model, frequency errors in the range of -800 kHz...800 kHz can theoretically be specified. The frequency f_0 of the local

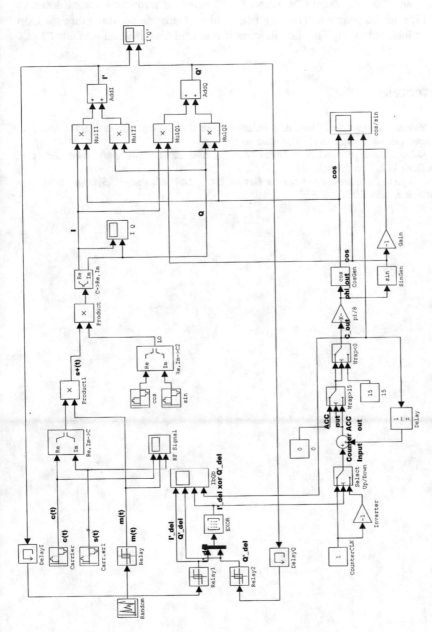

Fig. 8.9 Simulink model of modified Costas loop for BPSK using phasor rotator

oscillator is given by $f_0 = f_C - df$, with df = frequency error. Because Simulink does not support negative frequency values in the oscillator blocks, frequency errors larger than 400 kHz cannot be entered, but negative frequency errors down to -800 kHz are no problem. The simulations also clearly demonstrate that the loop acquires lock extremely fast, i.e., in at most one sampling interval (default 10 μs).

References

1. M. Vidmar, An improved BPSK demodulator for the 1.2 Mbit/s packet-radio RTX. http://lea.hamradio.si/ ~ s53mv/nbp/ax25/Bpskdem.pd
2. E. Roland, *Best, Phase-locked Loops, Design, Simulation, and Applications*, 6th edn. (McGraw-Hill, New York, 2007)
3. S.A. Tretter, *Communications System Design Using DSP Algorithms* (Springer Science + Business Media, 2008)

Chapter 9
Costas Loop for QPSK Using Phasor Rotator Circuit

9.1 Operating Principle

The Costas loop considered in this chapter is similar to the circuit discussed in Chap. 8 but is extended for operation with QPSK signals. The block diagram of this Costas loop is shown in Fig. 9.1.

The input signal is defined by

$$s(t) = m_1 \cos(\omega_1 t + \theta_1) - m_2 \sin(\omega_1 t + \theta_1)$$

where m_1 and m_2 are data signals that can have a value of +c or −c, where c is an arbitrary constant. In many cases, c = 1. ω_1 is the carrier radian frequency, and θ_1 is the zero phase. The local oscillator (labeled Osc) generates two output signals, a sine wave and a cosine wave having a constant radian frequency ω_2. s(t) is multiplied with the sine wave in the I branch and with the cosine wave in the Q branch. The output signals of the multipliers are lowpass filtered by filters LPF1 and LPF2 in order to remove the high frequency components at frequency $\omega_1 + \omega_2$. The output signal of lowpass filter LPF1 is then given by

$$I = m_1 \cos(\Delta\omega + \theta_e) - m_2 \sin(\Delta\omega + \theta_e)$$

and the output signal of LPF2 becomes

$$Q = m_2 \cos(\Delta\omega\, t + \theta_e) + m_1 \sin(\Delta\omega\, t + \theta_e)$$

Electronic supplementary material The online version of this chapter (doi:10.1007/978-3-319-72008-1_9) contains supplementary material, which is available to authorized users.

R. Best, *Costas Loops*, https://doi.org/10.1007/978-3-319-72008-1_9

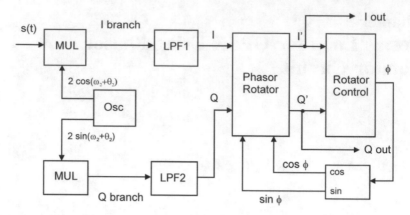

Fig. 9.1 Block diagram of Costas loop for QPSK using phasor rotator

with $\Delta\omega = \omega_1 - \omega_2$ and $\theta_e = \theta_1 - \theta_2$. In analogy to the circuit in Fig. 8.1, we define a phasor P by

$$P = I + jQ = (m_1 + j\, m_2)(\cos[\Delta\omega\, t + \theta_e] + j\, \sin[\Delta\omega\, t + \theta_e])$$
$$= (m_1 + j\, m_2)\exp(j[\Delta\omega t + \theta_e])$$

Phasor P rotates with radian frequency $\Delta\omega$, as in the loop of Fig. 8.1. The remaining blocks in Fig. 9.1 are the same as in the circuit of Fig. 8.1, but the schematic of the Rotator Control circuit is different, as will be explained below. The phasor rotator is required to cancel the rotation of phasor P by rotating P such that the rotated phasor P' settles at one of four positions as shown in Fig. 9.2. Assuming $c = 1$, the rotated phasor P' should be located either at P' = 1 + j (phase = 45°), or at P' = −1 + j (phase = 135°), or at P' = −1 − j (phase = 225°), or at P' = 1 − j (phase = 315°). When the loop has not yet acquired lock, the position of phasor P' can be anywhere, i.e., in any of the eight octants, which are defined in Fig. 9.2 and labeled O1...O8. Assume that P' is initially in octant O1. The phasor must then be rotated in positive direction (counterclockwise) to settle at an angle of 45°. When P' is in octant O2, however, it must be rotated in negative direction (clockwise) to settle at 45° as well. The phasor is always rotated to settle at the closest of the four phasor locations shown in Fig. 9.2.

As we have seen in Sect. 8.1, the phasor rotator can be considered as a rotating switch (cf. Fig. 8.2). The position of the switch is determined by the block labeled Rotator Control in Fig. 9.1. The block diagram of the Rotator Control is shown in Fig. 9.3. First the required direction of rotation must be determined. A logical signal $UP/\overline{DN}$ is generated, which controls the counting direction of an up/down counter. The counting direction depends on the octant where phasor P' is currently located. Table 9.1 shows the value of $UP/\overline{DN}$ in every one of the eight octants. $UP/\overline{DN}$ is a boolean function of three boolean variables I_b', Q_b', and C. I_b' is

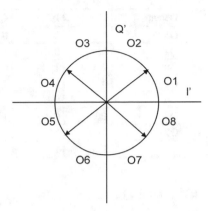

Fig. 9.2 Location of phasor P' with QPSK and definition of octants O1 ... O8

derived from signal I', cf. Fig. 9.3. I' is applied to the input of a comparator. I_b' is the output signal of that comparator. When I' is positive, I_b' is 1, otherwise it is 0. In analogy, Q_b' is a boolean signal which is 1 when Q' is positive and 0 when Q' is negative. The third boolean signal C is derived from the magnitude of I' and Q'. Note that the absolute value of I' is larger than the absolute value of Q' when phasor P' is in octant Q1. But the absolute value of I' is smaller than the absolute value of Q' when p' is in octant O2. Boolean variable C is used therefore to decide whether P' is in octant O1 or in octant O2. The boolean variable $UP/\overline{DN}$ is therefore a logical function of the three variables I_b', Q_b', and C.

As shown in Fig. 9.3, the signal $UP/\overline{DN}$ controls the counting direction of an up/down counter. When the signal is 1, the counter counts upwards, otherwise it counts downwards. The counter is clocked by a clock signal having frequency f_{clock}. In analogy to the circuit in Fig. 8.1, f_{clock} must be chosen an integer multiple of the symbol rate f_S, i.e., $f_{clock} = OS\, f_S$, with OS = oversampling factor. The

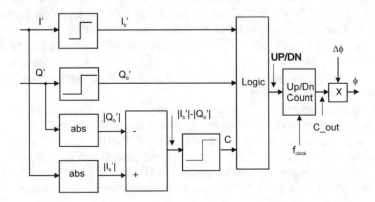

Fig. 9.3 Block diagram of phasor control circuit

Table 9.1 Truth table for the boolean function $UP/\overline{DN}$

Octant	C	$I_b{}'$	$Q_b{}'$	$UP/\overline{DN}$
O5	0	0	0	0
O4	0	0	1	1
O8	0	1	0	1
O1	0	1	1	0
O6	1	0	0	1
O3	1	0	1	0
O7	1	1	0	0
O2	1	1	1	1

content C_out of the up/down counter determines the position of the rotating switch (cf. Fig. 8.2).

The phasor rotator in Fig. 9.1 is identical with that of the Costas loop for BPSK described in Chap. 8, and its schematic is shown in Fig. 8.4.

9.2   Design Procedure for Costas Loop for QPSK Using Phasor Rotator

The design procedure for this type of Costas loop is very similar to that of the Costas loop discussed in Sect. 8.1. Three parameters have to be specified:

- the phase step $\Delta\phi$
- the oversampling factor OS
- the 3 dB corner frequency ω_{3dB} of the lowpass filters LPF1 and LPF2 (cf. Fig. 9.1).

Assume that the carrier frequency is $f_C = 400$ kHz and the symbol rate $f_S = 100'000$ bits/s. In case of the Costas loop for BPSK in Sect. 8.1, specifying a value $\Delta\phi = 2\pi/16$ (22.5°) was a good choice. When this value was chosen for the Costas loop for QPSK, simulations revealed that this value is too large. This resulted in instabilities of the loop; hence, it was necessary to reduce $\Delta\phi$. The choice $\Delta\phi = 2\pi/32$ was successful; thus, it is recommended to use it for this design.

The oversampling factor can be specified using Eq. (8.4). Choosing OS = 16 yields a pull-in range $\Delta f_P = 50$ kHz. When a larger pull-in range is desired, a larger value for OS can be specified.

Last we determine the 3 dB corner frequency of the lowpass filters. The modulating signal m(t) is a square wave function. The largest fundamental frequency of that signal is $f_S/2$, when a bit sequence of the form 01010101... is transmitted. ω_{3dB} must therefore be chosen markedly larger than $f_S/2$, but also smaller than twice the radian carrier frequency ω_C, which is $2\pi \cdot 400'000 = 2'512'000$ rad/s. A good choice would be to set the 3 dB corner frequency twice the symbol rate, i.e. $\omega_{3dB} = 2 \cdot 2\pi \cdot f_S = 1'256'000$ rad/s.

The transfer function of the lowpass filters is given by

$$H_{LPF}(s) = \frac{1}{1 + s/\omega_{3dB}}$$

When the lowpass filters are realized as digital filters, we must convert the transfer function $H_{LPF}(s)$ into the discrete transfer function $H_{LPF}(z)$. For best results, it is preferable to use the bilinear z transform [1]. Given an analog transfer function $H(s)$, this can be converted into a discrete transfer function $H(z)$ by replacing s by

$$s = \frac{2}{T}\frac{1 - z^{-1}}{1 + z^{-1}} \tag{9.1}$$

T is the sampling interval used for this digital filter, and $f_F = 1/T$ is the sampling frequency. Because the highest frequency component at the output of these filters is twice the carrier frequency 2 f_C; according to the Nyquist theorem, the sampling frequency f_F must be chosen at least twice that frequency, i.e., $f_F > 1.6$ MHz. Now the bilinear z transform has the property that the analog frequency range from 0 to ∞ is compressed to the digital frequency range from 0 to $f_{samp}/2$. To avoid undesired "shrinking" of the corner frequencies (ω_C and ω_3), these must be "pre-warped" accordingly, i.e., we must set

$$\omega_{3dB,p} = \frac{2}{T}tg\frac{\omega_{3dB}\;T}{2} \tag{9.2}$$

where $\omega_{3dB,p}$ is the prewarped corner frequency. Now we can apply the bilinear z transform to the transfer functions of the lowpass filters and get

$$H_{LPF}(z) = \frac{\left[1 + \frac{2}{\omega_{3dB,p}\;T}\right] + \left[1 - \frac{2}{\omega_{3dB,p}\;T}\right]z^{-1}}{1 + z^{-1}} \tag{9.3}$$

A Costas loop using these parameters will be presented in Sect. 9.3.

9.3   Simulating the Costas Loop for QPSK Using Phasor Rotator

The block diagram of a Costas loop for QPSK is shown in Fig. 9.4 (QPSK5.mdl). The model is very similar to the model of the Costas loop for BPSK in Fig. 8.6. The only differences are in the transmitter section where a QPSK signal is created and in the logical circuits (right top) where the $UP/\overline{DN}$ signal for the up/down counter is generated.

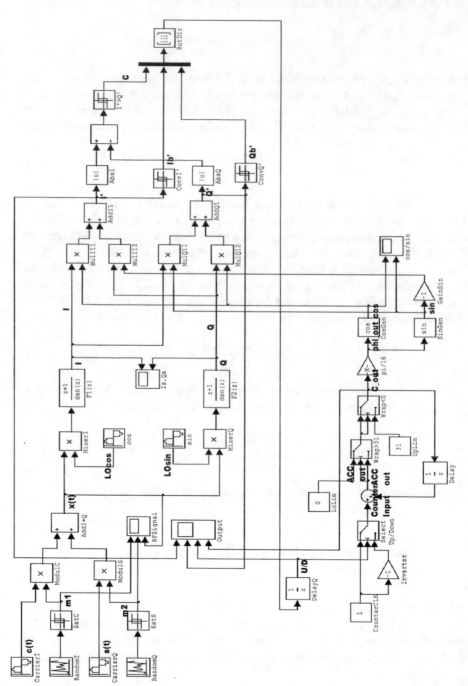

Fig. 9.4 Block diagram of Simulink model QPSK5.mdl

9.4 Modified Costas Loop for QPSK Using Phasor Rotator

In the preceding sections, we considered Costas loop for QPSK using phasor rotators that worked with a real input signal s(t). This type of Costas can realized, however, also as modified Costas loop, i.e., a Costas loop working with pre-envelope signals.

The block diagram of such a loop is shown in Fig. 9.5.

This circuit is almost identical with that of Fig. 9.1, with three minor differences:

1. The input signal s(t) is converted into a pre-envelope signal s + (t)
2. The local oscillator generates a complex carrier $\exp(-j[\omega_2 t + \theta_2])$
3. No lowpass filters are needed.

The input signal is given by

$$s(t) = m_1 \cos(\omega_1 t + \theta_1) - m_2 \sin(\omega_1 t + \theta_1)$$

where m_1 and m_2 are data signals that can have a value of +c or −c, where c is an arbitrary constant. In many cases, c = 1. The Hilbert transformed signal is then given by

$$\hat{s}(t) = m_1 \sin(\omega_1 t + \theta_1) + m_2 \cos(\omega_1 t + \theta_1)$$

and the pre-envelope signal then becomes

$$s^+ (t) = m_1 \cos(\omega_1 t + \theta_1) - m_2 \sin(\omega_1 t + \theta_1)$$
$$+ jm_1 \sin(\omega_1 t + \theta_1) + jm_2 \cos(\omega_1 t + \theta_1)$$

This can be rewritten as

$$s^+ (t) = (m_1 + jm_2)(\cos[\omega_1 t + \theta_1] + j \sin[\omega_1 t + \theta_1])$$
$$= (m_1 + jm_2) \exp(j[\omega_1 t + \theta_1])$$

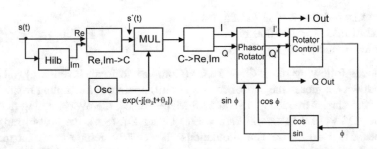

Fig. 9.5 Block diagram of modified Costas loop for QPSK using phasor rotator

Herein the term $m_1 + j \, m_2$ is called "complex envelope," and the term $exp(j\omega_1 t + \theta_1)$ is referred to as "complex carrier." The VCO generates another complex carrier given by

$$u_2(t) = \exp(-j[\omega_2 \, t + \theta_2])$$

The multiplier M_1 creates signal $u_m(t)$ that is given by

$$u_m(t) = (m_1 + j m_2) \exp(j[(\omega_1 - \omega_2) \, t + (\theta_1 - \theta_2)])$$

When the loop has acquired lock, $\omega_1 = \omega_2$, and $\theta_1 \approx \theta_2$, so we have

$$u_m(t) \approx (m_1 + j m_2)$$

The block C → Re, Im (Complex to Real, Imaginary) splits this signal into the real part I and the imaginary part Q. All other blocks of the circuit are identical with those of the previously discussed system (Fig. 9.1). The design of the modified Costas loop for QPSK is identical with the design of the previous system. Only two parameters have to be specified: the phase step $\Delta \phi$ and the oversampling factor OS.

9.5   Simulating the Modified Costas Loop for QPSK Using Phasor Rotator

A data transmission system using this modified Costas loop has been realized by model **QPSK6.mdl**. Its block diagram is shown in Fig. 9.6.

Most blocks are identical with those of the model in Fig. 9.4. The transmitter generates directly the pre-envelope signal $s^+(t)$. The local oscillator is built from a sine generator and a cosine generator that generate a fixed frequency signal. These two signals are combined by block Re,Im → C (Real, Image to Complex) to form the complex signal $exp(-j[\Delta\omega \, t + \theta_e])$. This signal is multiplied by the block labeled product. All remaining blocks are identical with those of the previous model (Fig. 8.6).

This model has been developed with the following default parameters:

- phase step $\Delta \phi = 2 \, \pi/32$ (11.25°)
- symbol rate $f_S = 100'000$ bits/s
- carrier frequency $f_C = 400$ kHz.

The blocks following block C → Re,Im (Complex to Real, Imaginary) (including the up/down counter in the phasor control circuit) operate at a sampling frequency of 32 times to carrier frequency, i.e., at 12.8 MHz. Hence, the oversampling factor OS becomes OS = 12.8 MHz/400 kHz = 128. Using Eq. (8.4), the pull-in range Δf_P becomes $\Delta f_P = 400$ kHz. The simulations also clearly demonstrate that the loop acquires lock extremely fast, i.e., in at most one sampling interval (default 10 μs).

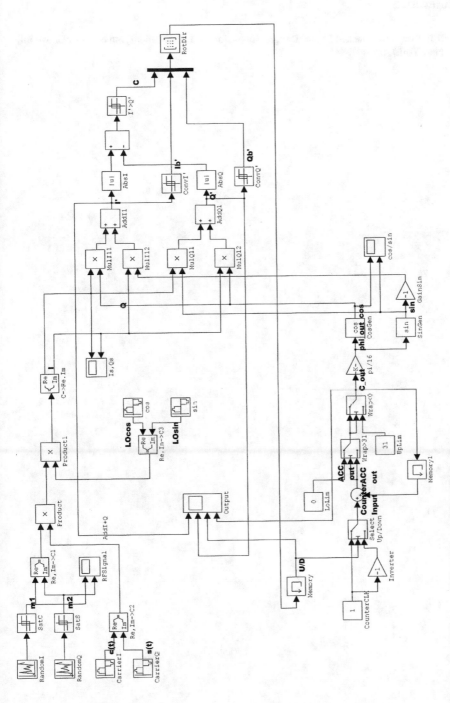

Fig. 9.6 Block diagram of Simulink model BPSK6.mdl

Reference

1. R.E. Best, Phase-locked Loops, Design, Simulation, and Applications, 6th edn. (McGraw-Hill, New York), pp. 441–446

Chapter 10
Costas Loop for Quadrature Amplitude Modulation

10.1 QAM Signal Generation

In the previous chapters, we considered Costas loops for BPSK, QPSK, and m-ary PSK. When using BPSK, one single bit is transmitted in every symbol interval. When the symbol rate is f_S, the information throughput of a BPSK link is f_S bits/s. With QPSK, two bits are transmitted in every symbol interval, i.e., the information throughput becomes $2 f_S$. More bits/s can be transmitted with m-ary PSK. When m is chosen 8, the information throughput is $3 f_S$ bits/s.

QAM can be considered an extension of QPSK. When using QPSK, two bit streams are transmitted at a time, where one bit stream builds the in-phase signal and the other bit stream the quadrature signal. Both bit streams are represented by binary signals, i.e., the modulating signals for the cosine and sine carriers can have amplitudes of 1 or −1 (respectively c or −c, where c is an arbitrary constant). With QAM, both amplitude and phase modulation are combined. This signifies that the modulating signals cannot take only values of 1 or −1, but more than two values, e.g., 3, −3, −1, 1, or 3. When both in-phase and quadrature signal can take four different values, their combination can take 16 different levels. This is shown in a so-called constellation diagram, as shown in Fig. 10.1.

This constellation is referred to as QAM_{16}. Using this modulation scheme, 4 bits can be transmitted in one symbol period. The information throughput can be further extended by using more than 16 constellation points, e.g., 64, 128, 256, or even more. With QAM_{256}, 8 bits are transmitted in one symbol period.

QAM is widely used today, for instance in digital TV or radio.

The generation of a QAM_{16} signal is shown in Fig. 10.2.

This kind of QAM is called *unfiltered* QAM. Both modulating signals m_1 and m_2 exhibit sharp transients at the start of the symbol intervals. The modulating signals

---

Electronic supplementary material The online version of this chapter (doi:10.1007/978-3-319-72008-1_10) contains supplementary material, which is available to authorized users.

Fig. 10.1 Constellation
diagram of QAM$_{16}$

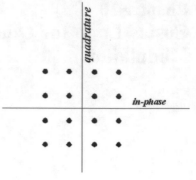

Fig. 10.2 Generation of
QAM$_{16}$ signal. **a** Modulating
signals m$_1$ and m$_2$ having
amplitudes of either -1.5,
-0.5, 0.5, or 1.5. **b** m$_1$ is
multiplied with cosine carrier,
m$_2$ is multiplied with sine
carrier

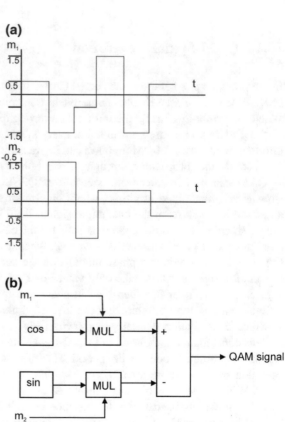

are similar to a square wave. When the polarity of the signal m$_1$ changes in every
symbol interval (i.e., when this signal looks like a sequence +−+−+−…), the
fundamental of that square wave like signal is half the symbol rate f$_S$. But because
such a signal contains also higher harmonics, the spectrum of signal m$_1$ is by far
larger than f$_S$/2 [1].

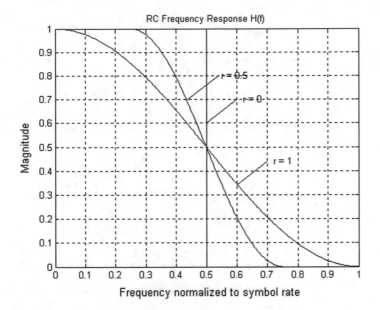

Fig. 10.3 Frequency response of Nyquist filters. Explanations cf. text

As Nyquist realized in 1928 [2], a receiver will be able to reconstruct the data signal if only the fundamental component of this square wave signal is transmitted, i.e., when the data signal is lowpass filtered with a corner frequency of $f_S/2$. A first idea would be to filter the m_1 and m_2 signals by a "brickwall filter," i.e., an ideal low-pass filter having gain 1 at frequencies below half the symbol rate $f_S/2$ and gain 0 elsewhere. The frequency response of the brickwall filter is shown in Fig. 10.3 by the curve labeled $r = 0$ (the meaning of r will be explained in the following).
The impulse response of the brickwall filter is

$$h(t) = \frac{\sin(\pi t/T)}{\pi t/T} \tag{10.1}$$

which is referred to as a *sinc* function. T is the symbol period. The duration of the impulse response is infinite, and it starts at $t = -\infty$; the filter delay is also infinite. The impulse response is plotted in Fig. 10.4 by the curve labeled $r = 0$. Of course such a filter cannot be realized. An approximation to the brickwall filter can be implemented by a FIR digital filter (FIR = finite impulse response) [3–5] to any desired level of accuracy, however because the impulse response decays slowly to 0, this leads to excessively long FIR filters, i.e., to filters with a large number of taps. Nevertheless, we will consider the impulse response in more detail because it exhibits a very useful property.

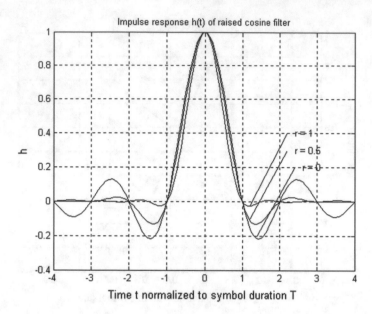

Fig. 10.4 Impulse response of the raised cosine filter for various values of (relative) excess bandwidth r

Every FIR filter causes the signal that is passed to be delayed, where the delay τ is given by

$$\tau = \frac{L-1}{2} T_F \qquad (10.2)$$

L is the length of the FIR filter and T_F is the sampling interval of the filter. In Fig. 10.4, the delay of the filter has been neglected; hence, its maximum (+1) occurs at time t = 0. Normally such a filter would be sampled at a frequency f_F which is an integer multiple of the symbol rate $f_S = 1/T$, where T is the symbol period. If we used a FIR filter of length 33, e.g., it would have a delay of 16 T_F; if we assume furthermore that the filter sampling frequency f_F is four times the symbol rate f_S, the filter would delay the signal by four symbol periods. For this FIR filter, the impulse response h(t) would no longer be symmetrical about t = 0, but rather about t = 4 T. For simplicity, the delay has been omitted in Fig. 10.3, i.e., we assume that the filter is a so-called *zero phase filter*, which causes no delay.

Note that the impulse response of the brickwall filter becomes 0 at t = T, 2T, 3T, etc. Now we remember that the impulse response is nothing more than the response of the filter to a delta function with amplitude 1, applied at t = 0 to its input. Sending a logical 1 therefore corresponds to applying a positive delta function (amplitude +1), and sending a logical 0 corresponds to applying a negative delta function (amplitude −1). Next we consider the case where a number of symbols it

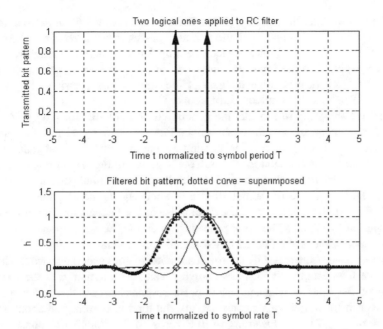

Fig. 10.5 Response of the raised cosine filter to a sequence of two binary ones applied to its input. Upper trace: a sequence of two delta functions corresponding to two binary 1's transmitted in succession. Lower trace: the solid curve (thin) whose maximum is at t = −T is the response of the filter to the delta function applied at t = −T, the solid curve whose maximum is at t = 0 is the response to the delta function applied at t = 0, and the thick (dotted) curve is the superposition of these two responses

passed through the filter in succession. If two succeeding 1's are supplied, the first logical 1 will be represented by a delta function with amplitude +1 applied at t = −T, the second logical 1 by another delta function with amplitude +1 applied at t = 0. This situation has been plotted in Fig. 10.5. The upper trace shows the two delta functions (marked by arrows), the lower trace the corresponding impulse responses (solid curves). The zeros of the impulse responses are marked by circles. The superposition of the two impulse responses is shown by the dotted curve (thick). In order to recover the data, the receiver would have to sample the filtered signal exactly at time t = 0, T, 2T, etc. As can be seen from Fig. 10.4, the amplitude of the combined signal at t = 0 (marked by a square) depends uniquely on the impulse applied at t = 0; the impulse response caused by the second delta function is zero at this time. The same holds for the combined signal at t = −T. Its amplitude at t = −T (marked by a square) uniquely depends on the amplitude of the second delta function and does not contain any contribution from the first delta function. In other words, the impulse responses caused by various delta functions do *not interfere*, i.e., when using the brickwall filter, there will no *inter symbol interference* (ISI). To recover the data signal without error, the applied filter must be chosen such that no ISI can occur. (It is easy to demonstrate how ISI can be created:

assume that we filter the data signal by a Butterworth lowpass filter, e.g., its impulse response will be a damped oscillation. It also passes through zero, but these zeros are not at $t = 0$, T, 2T, etc., hence the impulse responses coming from delta functions applied to the input will interfere, or in other words, an output sample taken at $t = kT$ (k = positive integer) is the sum of the contributions of many symbols. This cannot be tolerated, of course.)

Next we are going to consider the generation of a Nyquist-filtered QAM signal. Two data signals are given, as shown in Fig. 10.2a, a signal m_1, the in-phase component, and a signal m_2, the quadrature component. These signals are first converted to a series of delta functions. This is shown in Fig. 10.6 for the m_1 signal. The delta pulses are created at integer multiples of the symbol period, i.e., at $t = 0$, T, 2 T. The amplitude of the delta pulse at $t = 0$ equals the amplitude of the m_1 signal at $t = 0$, the amplitude of the delta pulse at $t = t$ equals the amplitude of the m_1 signal at $t = T$, etc. The same procedure is applied to signal m_2.

The generation of the filtered QAM signal is represented by Fig. 10.7.

As we have seen a useful approximation to a brickwall filter would lead to excessive filter length. Nyquist has shown a valuable alternative: if the transition region of the filter is widened, its impulse response decays faster toward zero. Moreover, if the amplitude response $H(f)$ of the filter is symmetrical about half the symbol rate ($f_S/2$), the locations of the zeros of its impulse response remain unchanged. Filters having this property are commonly referred to as "*Nyquist filters.*" One possible realization of the Nyquist filter is the ***raised cosine filter*** (RCF). Its transition region (the region between passband and stopband is shaped by a "raised" cosine function, i.e., by a cosine wave which stands "on a piedestal." The width of the transition band is determined by the parameter r which is called excess bandwidth. If the frequency corresponding to half the symbol rate is denoted $f_S/2$, the transition band starts at $f = (1 - r) f_S/2$ and ends at $(1 + r) f_S/2$. The amplitude response $H(s)$ of the RCF has been plotted in Fig. 10.4 for three values of r, $r = 0$, 0.5, and 1 (the case $r = 0$ applies for the brickwall filter). Note that the frequency response is symmetrical about half the symbol rate. Mathematically, the frequency response of the RCF is given by

Fig. 10.6 Converting the continuous signal m_1 into a series of delta functions

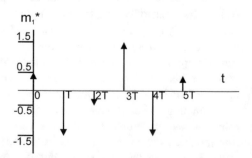

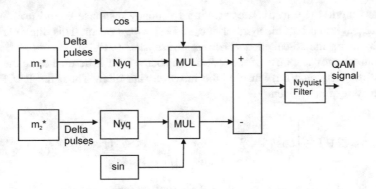

Fig. 10.7   Generation of Nyquist-filtered QAM signal

$$H_{RCF}(f) = \left| \begin{array}{ccc} 1 & for & |f| < \frac{1-r}{2T} \\ \cos^2 \frac{\pi}{4} \frac{2T|f|+r-1}{r} & for & \frac{1-r}{2T} \leq |f| \\ 0 & for & |f| > \frac{1+r}{2T} \end{array} \right. \tag{10.3}$$

Figure 10.4 shows the impulse response of the RCF for r = 0, 0.5, and 1. The larger r is chosen, the faster the decay becomes. To economize bandwidth, however, large values of r must be avoided. In practice, values of r in the range 0.15...0.35 are customary. For completeness, we also give the expression for the impulse response $h_{RCF}(t)$ of the RCF [3].

$$h_{RCF}(t) = \frac{\sin(\pi t/T) \cdot \cos(\pi r t/T)}{(\pi t/T) \cdot \left[1 - \left(\frac{2rt}{T} \right)^2 \right]} \tag{10.4}$$

Designing of an FIR raised cosine filter is an easy task: Eq. (10.4) can be used to compute the filter coefficients. Care must be taken, however, since there exist values of t where both numerator and denominator become 0. This happens if the expression in the square brackets of the denominator becomes 0, explicitly for t/T = 1/(2r). For the same value of t, the cosine term also becomes 0. This singularity is removed by replacing both numerator and denominator by their derivatives; mathematically, this is called L'Hôpital's rule. The computation is made even easier if MATLAB's Signal Processing Toolbox is available. It contains a function called FIRRCOS that calculates the coefficients of the RCF.

So far the raised cosine filter seems to offer the optimum solution for Nyquist filtering because it completely suppresses ISI. As we will see in the following sections, the receiver also needs a lowpass filter for different reasons. It will show up that we cannot use another raised cosine filter at the receiver: when doing so, the data signal would have to pass through two cascaded RCF's: one in the transmitter and one in the receiver. The overall frequency response then would be the RCF transfer function **squared**! The resulting frequency response would no longer be symmetrical about half the symbol rate, and ISI would occur. This dilemma can be fixed by using the so-called *root raised cosine filter* (RRCF).

The frequency response of the root raised cosine filter is defined to be the square root of the frequency response of the raised cosine filter. Thus, the RRCF has a frequency response given

$$H_{RRCF}(f) = \begin{vmatrix} 1 & for & |f| & < \frac{1-r}{2T} \\ \cos\frac{\pi}{4}\frac{2T|f|+r-1}{r} & for & \frac{1-r}{2T} & \leq & |f| & \leq & \frac{1+r}{2T} \\ 0 & for & |f| & > & \frac{1+r}{2T} \end{vmatrix} \qquad (10.5)$$

The impulse response $h_{RRCF}(t)$ of the RRCF is given by [3]

$$h_{RRCF}(t) = \frac{\sin\left[\frac{\pi t}{T}(1-r)\right] + \frac{4rt}{T}\cos\left[\frac{\pi t}{T}(1+r)\right]}{\frac{\pi t}{T}\left[1 - \left(\frac{4rt}{T}\right)^2\right]} \qquad (10.6)$$

Similar to the impulse response of the RCF, the impulse response given by Eq. (10.6) also has singularities. For some values of time t, h(t) becomes a division 0/0. In analogy, this problem is mastered by applying L'Hôpital's rule and replacing numerator and denominator by its derivatives.

The frequency response of the RRCF is no longer symmetrical about half the symbol rate; hence, it leads to ISI which, of course, is not desired. But if the receiver also contains an identical RRCF, the overall frequency response is identical with the RCF, hence after the demodulator, ISI becomes zero again.

The Costas loop considered in the following will use RRCF's for Nyquist filtering, and it is assumed, that another RRCF is used in the transmitter.

10.2   Structure of a Costas Loop for QAM

A block diagram of a Costas loop for QAM is depicted in Fig. 10.8.

10.2.1   Nyquist Filtering of Input Signal S(T) (RRCF1 and RRCF2)

The structure is similar to the Costas loop for QPSK, as shown in Fig. 4.1. In the I branch, the incoming QAM signal s(t) is multiplied with a cosine signal generated by local oscillator VCO. In analogy, s(t) is multiplied in the Q branch with a sine signal. When the loop is locked onto the carrier in both frequency and phase, signals I and Q are identical with the data signals m_1 and m_2, filtered with a raised

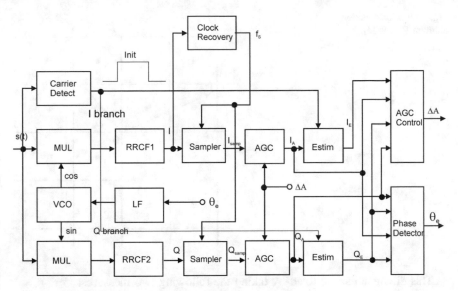

Fig. 10.8 Block diagram of a Costas loop for QAM

cosine filter. To recover the unfiltered signals m_1 and m_2, the signals I and Q must be sampled at the correct instants of time, i.e., at times $t = 0, T, 2T \ldots$ (assuming that the filters are zero phase filters—in reality, the signals are delayed by twice the delay of one root raised cosine filter). The sampled I and Q signals are labeled I_{samp} and Q_{samp}, respectively.

10.2.2   Automatic Gain Control (AGC)

Assume for the moment that QAM_{16} is used in this Costas loop. When no phase error and no amplitude error is present (more about amplitude errors later), the phasor built from I_{samp} and Q_{samp} would coincide exactly with one of the constellation points shown in Fig. 10.9. Because the QAM signal is transmitted over a link, however, the amplitude of the received QAM signal s(t) can deviate from the amplitude of the transmitter output signal. When the signal is sent over a long cable, e.g., the signal can be attenuated. When repeaters are used within the link (e.g., with wireless communication), the amplitude of s(t) can also be larger than the transmitter output signal. This situation is illustrated by phasor P_1 in Fig. 10.9; this phasor has the coordinates $I_A = 0.8$, $Q_A = 0.95$, i.e., $P_1 = (0.8, 0.95)$. When the gain of the link is less than 1 (attenuation), the actual value of the current phasor could be (1.5, 1.5), but when the gain of the link is larger than 1 (amplification), the actual value of the phasor could be (0.5, 0.5).

Fig. 10.9 Constellation
diagram for QAM$_{16}$

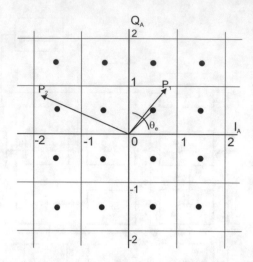

This dilemma can be fixed by taking the following two measures:

1. The transmission of symbols starts with a preamble. The preamble can be given
   by a series of equal symbols, e.g., P = (0.5, 0.5). The receiver is equipped with a
   carrier detect circuit (cf. Fig. 10.8). As soon the receiver detects a carrier, the
   carrier detect block generates an initialization pulse labeled "Init" in Fig. 10.8.
   When the symbol rate is 100'000 symbols/s and 20 symbols are transmitted in
   the preamble, the duration of the Init pulse will be 20 · 10 μs = 200 μs. The Init
   pulse is applied to the control input of the blocks labeled "Estim." The Estim
   blocks estimate the amplitude of both I and Q signals. During the preamble, the
   Estim block in the I branch generates an output signal I_E = 0.5, and the Estim
   block in the Q branch generates an output signal Q_E = 0.5.

2. An automatic gain control system is added to the Costas loop, cf. the blocks labeled
   AGC in Fig. 10.8. The AGC blocks are variable gain amplifiers. Their gain is
   controled by signal ΔA. ΔA is generated by the block labeled AGC control. This
   block computes the magnitude M_A of the actual symbol (I_A, Q_A) from

$$M_A = +\sqrt{I_A^2 + Q_A^2} \tag{10.7a}$$

and the magnitude of the estimated phasor (I_E, Q_E) from

$$M_E = +\sqrt{I_E^2 + Q_E^2} \tag{10.7b}$$

The amplitude error ΔA is now calculated from

$$\Delta A = M_A - M_E \tag{10.7c}$$

Signal ΔA is applied to the control inputs of the AGC blocks. The gain of the AGC amplifiers is successively adjusted such that finally the amplitude error becomes 0 on average.

10.2.3   Phase Detector

It should be noted that at start of the initialization period the Costas loop is normally not yet locked in both frequency and phase onto the carrier frequency; hence, there will also be phase errors θ_e. This situation is sketched in Fig. 10.9 by phasor P_2. When the local oscillator (VCO) is locked to the frequency of the carrier, this phase error would be a constant. But when the frequency of the VCO deviates from the carrier frequency, the phasor P_2 would rotate with a frequency given by the difference between local oscillator frequency and carrier frequency, as we have seen in previous chapters.

The block labeled "Phase Detector" computes the phase error of current phasor (I_A, Q_A). During the preamble interval, the phase error is computed from

$$\theta_e = phase(I_A, Q_A) - phase(I_E, Q_E) \qquad (10.8)$$

The phase of a phasor is often computed by the arctangent function. MATLAB offers two different arctangent functions, atan and atan2. atan2 is preferred because it computes phases within a range of $-180...180°$, whereas atan computes the phase only within a range of $-90...90°$. The atan2 function is visualized in Fig. 10.10. When the phasor (I_A, Q_A) is in quadrant Q1, the phase can vary within a range of $0...90°$. When the phasor is in quadrant Q2, the phase is in the range $90...180°$. In this quadrant, the atan function would deliver phase argument in the range from 0 to $-90°$. In quadrant Q3, the phase is in the range $-180...-90°$, and in quandrant Q4, the phase is in the range $-90...0°$. The atan2 function is preferred because it delivers argument over a larger range. We will see later that the phase error can exceed $180°$ during the acquisition process and can even take values larger than a

Fig. 10.10 Definition of quadrants and range of the atan2 function in MATLAB

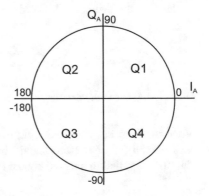

multiple of 2π or less than a multiple of -2π. To cover a range larger than $-180°$...
$180°$, special algorithms will be developed.

10.2.4 Estimator (Estim)

We have seen that during the preamble the Estim blocks delivers the actual
amplitude of the phasors (I_E, Q_E) transmitted during that interval of time. When a
series of identical phasors (0.5, 0.5) is sent during the preamble, the Estim block in
the I branch generates an output signal $I_E = 0.5$, and the Estim block in the Q
branch generates an output $Q_E = 0.5$. When the preamble is over, the loop has
acquired lock, and the gain of the AGC blocks has been set in order to receive the
correct amplitudes of phasor (I_A, Q_A). Because phase and amplitude errors are not
exactly zero due to broadband noise superimposed to signal s(t), the sampled
amplitudes I_A, Q_A can slightly deviate from the actual amplitudes of the transmitted
symbol. This is shown in Fig. 10.9 by phasor P_1. Assume that a phasor P = (0.5,
0.5) has been transmitted. Due to amplitude and phase errors, the received symbol
P1 deviates from this value. The correct amplitudes are estimated then using a most
likelihood algorithm. The constellation plane I_A, Q_A is subdivided into squares as
shown in the figure, and the constellation points are in the center of these squares.
To determine the correct amplitudes of a received phasor, the Estim blocks takes
that constellation point which is closest to the received on. This is done by a
quantizing the amplitudes I_A, Q_A as follows:

1. The Estim block in the I branch is a quantizer for signal I_A, i.e., the estimated
   signal I_E is obtained from

$$I_E = quant(I_A)$$

   I_E is set -1.5 when I_A is less than -1, and I_E is set -0.5 when I_A is in the range
   $-1...0$, and I_E is set 0.5 when I_A is in the range $0...1$, and I_E is set 1.5 when I_A is
   greater than 1.
2. The same procedure is performed in the Estim block in the Q branch. This block
   delivers the estimate Q_E for the received Q_A signal by the operation

$$Q_E = quant(Q_A)$$

10.2.5 Clock Recovery

When the loop has acquired lock, the signal I at the output of the RRCF1 filter in
the I branch (cf. Fig. 10.8) should be identical with the series of delta pulses (cf.
Fig. 10.6) applied to a raised cosine filter (RCF). When the gain of the transmission

link is not exactly 1, this is corrected by the AGC system. In Fig. 10.4, the impulse response of the RCF has been shown. Assuming that the filter is a zero phase filter, the impulse response of the delta pulse m_1^* applied at t = 0 has exactly the value m_1^* at t = 0. The amplitude of the delta pulse applied at t = T has exactly the value m_1^* at time t = T, etc. It is therefore extremely important to sample the output signal of RRCF1 at the correct times t = 0, T, 2T ... The block **Clock Recovery** is used to recover a clock signal, i.e., a series of strobe signals having the frequency f_S (Symbol rate). These strobes sample signal I at the correct times, and the sampled signal I_{samp} is obtained from the output of block **Sampler**.

There are many algorithms for clock recovery. Here we are using a method developed by Gardner [6, 7], the so-called **prefilter method**. The operating principle of block **Clock Recovery** is explained by the block diagram in Fig. 10.11.

The frequency response $H_{pre}(f)$ of the prefilter is derived from the frequency response $H_{RCF}(f)$ of the raised cosine filter (cf. Eq. 10.3):

$$H_{Pr\,e}(f) = H_{RCF}(f - f_S) + H_{RCF}(f + f_S) \tag{10.9}$$

with f_S = symbol rate. The frequency response of the RCF is plotted once again in Fig. 10.12, upper trace. The frequency response of the prefilter (middle trace) consists of two parts, the frequency response of the RCF shifted left by f_S and the frequency response of the RCF shifted right by f_S. Now we remember that the data signal m_1^* (a series of delta pulses) has passed through a RRCF in the transmitter, as shown in Fig. 10.7. After transmission through a link, this signal has passed another RRCF in the receiver (cf. RRCF1 in Fig. 10.8). This signal is labeled I. Transmission through a cascade of two RRCF's is equivalent to transmission through one RCF filter, as explained in Sect. 10.1. In block Clock Recovery, I passes through the prefilter, as shown in Fig. 10.11. The data signal m_1^* has therefore passed the cascade of an RCF and the prefilter. The overall frequency transfer function H_{tot} for the data signal in the I branch is therefore given by

$$H_{tot}(f) = H_{RCF}(f) \cdot H_{Pr\,e}(f) \tag{10.10}$$

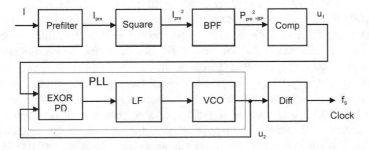

Fig. 10.11 Block diagram of prefilter algorithm

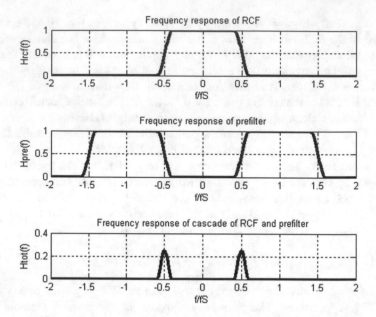

Fig. 10.12 Frequency response of RCF (upper trace), prefilter (middle trace), and cascade of RCF and prefilter (lower trace)

$H_{tot}(f)$ is plotted in the bottom trace of Fig. 10.12. We recognize that this is the frequency response of a bandpass filter having a center (resonance) frequency $f_S/2$. When a delta pulse is applied at $t = 0$ to a filter with frequency response $H_{tot}(f)$, the output of filter $H_{tot}(t)$ (I_{Pre} in Fig. 10.11) is a damped cosine wave with a frequency $f_S/2$. Assuming that all filters involved are zero phase filters, it can be shown that I_{Pre} has zero as $t = 0.5$ T, 1.5 T, 2.5 T ... (With an actual filter, the impulse response will be the same, but delayed by the sum of delays of all filters involved, hence the position of the zeros will be $t = 0.5$ T $+ T_d$, 1.5 T $+ T_d$, 2.5 T $+ T_d$ ...., with T_d = total delay of filters.) When another delta pulse is applied at $t + T$, then the impulse response of H_{tot} to that pulse has, assuming again zero phase filters, zeros at $t = 0.5$ T, 1.5 T, 2.5 T ... Hence, the superposition of responses onto delta pulses applied at $t = 0$, T, 2T, etc., will still have zeros at $t = 0.5$ T, 1.5 T ... This makes it possible to derive a clock signal by locating the positions of the zeros.

The first trace in Fig. 10.13 shows the prefilter output signal. The waveforms in this figure are taken from a simulation performed with a Simulink model. This model will be described in detail in Sect. 10.4. In this model, the symbol rate f_S has been chosen 100'000 symbols/s; hence, we have T = 10 μs. It is clearly seen that the zeros occur every 10 μs. The prefilter signal has a frequency of $f_S/2 = 50$ kHz. Following Gardners algorithm, the prefilter output signal is first squared, cf. second trace in Fig. 10.13. The squared signal has now a frequency of $f_S = 100$ kHz and has again zeros separated by 10 μs. To remove the dc term of the squared signal, this signal is filtered by a bandpass filter having a center frequency of 100 kHz. The

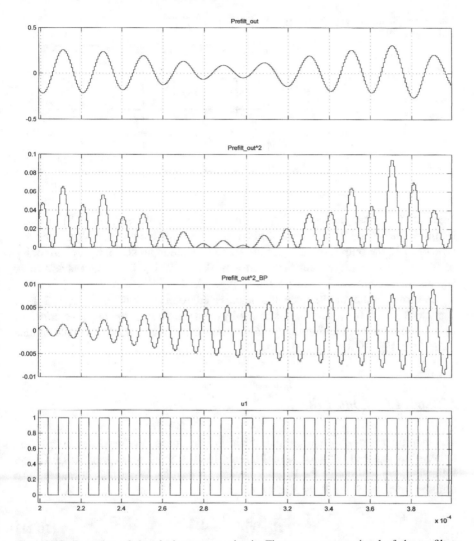

Fig. 10.13 Operation of the clock recovery circuit. First trace: output signal of the prefilter, second trace: output signal of the prefilter squared, third trace: output signal of the prefilter squared and bandpass filtered, fourth trace: output of comparator, the signal is 1 when the bandpass-filtered output signal is positive and 0 when the bandpass-filtered output signal is negative

bandpass-filtered signal is shown in the third trace. The zero transitions of that signal are now at $t = T/4$, $3 T/4$, $5T/4$ ... The bandpass filtered signal is applied to a comparator delivering a logical signal. The output signal is 1 when the bandpass filtered signal is positive and 0 when that signal is negative. This logical signal is displayed in the fourth trace and is labeled u_1. Signal u_1 is applied to the reference input of a PLL (cf. Fig. 10.11) that operates with square wave signals. The phase detector of this PLL is realized by an EXOR gate [3, 8, 9]. As usual, the output

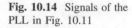

Fig. 10.14 Signals of the
PLL in Fig. 10.11

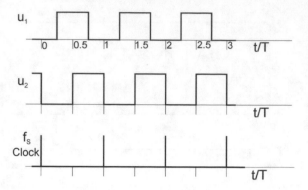

signal of the phase detector is applied to the input of a loop filter (LF), and the
output signal of the loop filter is applied to the control input of the VCO. For this
type of PLL, the output signal u_2 of the VCO lags the input signal by a phase delay
of 90°. u_2 has therefore transitions at t = 0, 0.5 T, T, 1.5 T ... The waveform of
signals u_1, u_2, and clock is displayed in Fig. 10.14.

As shown in the middle trace, the negative edges occur at times t = 0, T, 2 T ...
The clock signal f_S is therefore obtained by "differentiating" the negative edges of
u_2, cf. bottom trace as in Fig. 10.14. This is the strobe signal used to sample signals
I and Q in Fig. 10.8.

10.2.6   LF (Loop Filter)

The loop filter is mostly realized like the loop filters used in other Costas loops, i.e.,
either as a lag-lead filter or as a PI filter (proportional + integral filter). In the
following, we will prefer the PI filter because it enables larger pull-in ranges. The
transfer function of the PI is given by

$$H_{LF}(s) = \frac{1 + s\tau_2}{s\tau_1} \qquad (10.11)$$

with τ_1, τ_2 = time constants. In most applications, the loop filter is realized by a
digital filter; hence, the transfer function $H_{LF}(s)$ must be converted to the discrete
time transfer function $H_{LF}(z)$. This will be discussed in more detail in Sect. 10.3.

10.2.7   Voltage-Controlled Oscillator (VCO)

The VCO is a circuit delivering simultaneously a cosine and a sine output signal. As
we have seen in previous chapters, the phase transfer function $H_{VCO}(s)$ is given by

$$H_{VCO}(s) = \frac{K_0}{s} \qquad (10.12)$$

with K_0 = VCO gain. In most applications, the VCO is realized by a digital filter; hence, the transfer function $H_{VCO}(s)$ must be converted to the discrete time transfer function $H_{VCO}(z)$. This will be discussed in more detail in Sect. 10.3.

10.3 Design Procedure for Costas Loop for QAM

Because the number of functions blocks in a Costas loop for QAM is considerably higher than in case of Costas loops for BPSK or QPSK, the design procedure for this type of Costas loop is much more complex and time-consuming. This relates specially to the large number of special filters such as RRCFs and prefilters. In the following, the mathematical background for the design of those function block is presented. All blocks discussed below are shown in Fig. 10.8.

10.3.1 Blocks RRCF1 and RRCF2 (Root Raised Cosine Filters)

The RRCFs are realized by Finite Impulse Response filters (FIR). The impulse response $h_{RRCF}(t)$ is given by Eq. (10.6). This is a continuous function of time, and its duration is from $t = -\infty$ to $t = \infty$. To get a realizable digital filter, the duration must be truncated, and the function $h_{RRCF}(t)$ must be discretized, i.e., transformed to a series of discrete samples $h_{RRCF}*$. We remember that in the system in Fig. 10.8 there are two RRCFs in series, one in the transmitter and one in the receiver (Costas loop). The impulse response of the cascaded RRCFs is identical with that of the RCF, as shown in Fig. 10.4. To get an idea what should be about the duration of the truncated impulse response, we recognize that the impulse response of the RCF (e.g., r = 0.5) is almost zero for times less than $-4T$ and greater than 4T. It is therefore reasonable to truncate the impulse response h_{RRCF} to a time range from $-nT < t < nT$, where n is an integer in the order of 4. To get a discrete impulse response, h_{RRCF} must be sampled by a frequency f_f larger than the symbol frequency $f_S = 1/T$, with T = symbol duration. Let f_f be $f_f = OS \cdot f_S$, where OS is an oversampling factor. Because the output signal of the RRCF is applied to the prefilter and the output signal of the prefilter is used to determine the zero transitions of the signal I, OS must be chosen large enough to determine the time instants of the zero transitions with sufficient accuracy. It is therefore adequate to choose OS = 16 or larger. According to the theory of digital filters, the discrete impulse response $h_{RRCF}*$ of the FIR filter is calculated from the value of h_{RRCF} of the continuous impulse response multiplied by the sampling interval $T_f = 1/f_f$, i.e.,

$h_{RRCF}* = h_{RRCF} \cdot T_f$ [3, 5]. To implement the FIR filter, we first calculate a time vector t[i], i.e., a vector built from the sampling instants of the FIR filter. This yields

$$t[i] = (i + \varepsilon)\frac{T}{OS}, \quad i = -n\,OS\ldots n\,OS \tag{10.13}$$

We remember that when computing the impulse response of the continuous RRCF there are time instants t where the $h_{RCF}(t)$ becomes a division 0/0. To avoid this, we add a little time offset ε to index i. When ε is chosen very small, e.g., 0.001, this effect is negligible, but we avoid the division 0/0. Now the discrete impulse response of the RRCF becomes a vector $h_{RRCF}[i]*$

$$h_{RRCF}[i]* = \frac{\sin\left[\frac{\pi\,t[i]}{T}(1-r)\right] + \frac{4rt[i]}{T}\cos\left[\frac{\pi\,t[i]}{T}(1+r)\right]}{\frac{\pi\,t[i]}{T}\left[1 - \left(\frac{4\,r\,t[i]}{T}\right)^2\right]} \cdot \frac{T}{OS} \tag{10.14}$$

$$i = -n\,OS\ldots n\,OS$$

This vector yields the values of the numerator coefficients of the FIR filter. The denominator coefficients do not exist, because the discrete transfer function of a FIR filter is by definition

$$H(z) = a_0 + a_1 z^{-1} + a_2 z^{-2} + \ldots$$

with a_i = filter coefficients.

10.3.2   Clock Recovery

The block diagram of the clock recovery system is shown in Fig. 10.11. We will first develop the prefilter. This filter will also be realized as a FIR filter; hence, we must know the filter coefficients. The frequency response $H_{pre}(f)$ of the prefilter (Eq. 10.9) is derived from the frequency response of the RCF (Eq. 10.3), because it consists of two shifted replicas of $H_{RCF}(f)$, i.e., we have

$$H_{Pre}(f) = H_{RCF}(f - f_S) + H_{RCF}(f + f_S)$$

When the impulse response h(t) of a filter having frequency response H(f) is known, the impulse response of a filter having frequency response $H(f - f_S)$ is given, according to the shift theorem of the Fourier transform, by

$$h_{shifted}(t) = h(t)\exp(j\,2\pi f_S)$$

Consequently, the impulse response of the prefilter is given by

$$h_{Pre}(t) = h_{RCF}(t)(\exp[j\,2\pi f_S] + \exp[j\,2\pi f_S]) = 2h_{RCF}(t)\cos(2\pi f_S) \qquad (10.15)$$

To realize the filter, we can apply the same procedure as done for the RRCF in the last paragraph. The impulse response $h_{PRE}(f)$ is first truncated within a range of time $-n\,T < t < n\,T$, where T is the symbol interval and n is an small integer, e.g., n = 4. Moreover, the impulse response must be converted to discrete series $h_{Pre}^*(t$ [i]) where I is an index ranging from $-n$ OS ... n OS. Again, OS is an oversampling factor. The impulse response must be sampled at a frequency larger than the symbol rate f_S, i.e., the sampling frequency f_f is

$$f_f = OS \cdot f_S$$

OS must be chosen large enough to get a sufficient time resolution; a value of OS = 16 is adequate in most cases. As we did with the RCF, a time vector is computed first (Eq. 10.13):

$$t[i] = (i + \varepsilon)\frac{T}{OS}, \quad i = -n\,OS \ldots n\,OS$$

The discrete impulse response of the prefilter is then given by

$$h_{RCF}(t[i]) = \frac{\sin(\pi\,t[i]/T) \cdot \cos(\pi\,r\,t[i]/T)}{(\pi\,t[i]/T) \cdot \left[1 - \left(\frac{2\,r\,t[i]}{T}\right)^2\right]} \cdot \frac{T}{OS} \qquad (10.16)$$

These values are identical with the filter coefficients of the numerator of the FIR filter.

The bandpass filter in Fig. 10.11 can be realized by an Infinite Impulse Response (IIR) filter [4, 5]. Designing bandpass filters is very simple with MATLAB. MATLAB has a function "butter" that designs an IIR Butterworth filter. The coefficients [B, A] are obtained by calling

$$[B, A] = \text{butter}(n, Wn)$$

where n is the order of the filter. To get a bandpass filter, Wn must be specified as a vector Wn = [W1, W2], where W1 is the lower 3 dB corner frequency and W2 is the upper 3 dB corner frequency of the filter. Both W1 and W2 must be related to the Nyquist frequency, i.e., to half the sampling frequency f_f. Thus, we obtain the bandpass filter by specifying

$$[B, A] = butter\left(2, \left[\frac{f_S - \Delta f}{f_f/2}, \frac{f_S + \Delta f}{f_f/2}\right]\right)$$

with Δf = one-sided bandwidth of the bandpass filter. When f_S is the resonant frequency of the bandpass filter, Δf can be chosen $\Delta f = 0.01\ f_S$. Vectors B and A represent the filter coefficients: B contains the coefficients of the numerator, A the coefficients of the denominator. When a bandpass filter with n = 2 is specified, MATLAB creates a filter with order 2 n = 4.

Next the PLL system of Fig. 10.11 must be designed. This circuit operates with binary signals exclusively. It consists of a phase detector realized with an EXOR gate, a loop filter realized as a PI filter (proportional + integral), and a VCO. The transfer function of the phase detector is [3]

$$H_{PD}(s) = \frac{1}{\pi} = K_d \tag{10.17}$$

with Kd = phase detector gain. The transfer function of the loop filter is given by

$$H_{LF}(s) = \frac{1 + s\tau_2}{s\tau_1} \tag{10.18}$$

with τ_1, τ_2 = time constants, and the transfer function of the VCO becomes

$$H_{VCO}(s) = \frac{K_0}{s} \tag{10.19}$$

with K_0 = VCO gain. The PLL operates at a center frequency $f_0 = f_S$, where f_S is the symbol rate. The parameters of the PLL are determined in the same way we already used in the design of Costas loops for BPSK and QPSK. First the open loop gain $G_{OL}(\omega)$ is considered, cf. Fig. 10.15. Using Eqs. (10.17)–(10.19), the open loop gain $G_{OL}(\omega)$ is given by

$$G_{OL}(s) = \frac{K_0 K_d}{s} \cdot \frac{1 + s\tau_2}{s\tau_1} \tag{10.20}$$

Fig. 10.15 Open loop gain GOL(ω) of the PLL

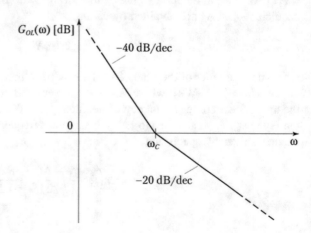

In this figure, the magnitude of $G_{OL}(\omega)$ is plotted (Bode diagram). ω_C is the corner frequency $1/\tau_2$ of the loop filter. Below that frequency the gain rolls off at a slope of -40 dB/dec and above it rolls off at a slope of -20 dB/dec. In this design, ω_C has been chosen to be identical with the transit frequency ω_T of the loop, which is the frequency where the open loop gain is 1. With this choice, the phase margin of the loop becomes $45°$, which is sufficient for stability. Given the center frequency f_0 of the loop, the transit frequency is usually chosen to be about $0.05 \ldots 0.1$ the center frequency. In our example, it is adequate to specify $\omega_T = 0.1 \, \omega_0 = 0.1 \cdot 2\,\pi \, f_S$, where ω_0 is the radian center frequency of the loop. Given the transit frequency ω_T, the time constant τ_2 becomes

$$\tau_2 = \frac{1}{\omega_T} = \frac{1}{0.1 \cdot 2\pi f_S} \tag{10.21}$$

The remaining parameters K_0 and τ_1 must now be chosen such that the open loop gain becomes 1 at $\omega = \omega_T$, i.e.,

$$1 = \frac{K_0 \, K_d}{\omega_T^2 \, \tau_1}$$

Because this equation is overdetermined, one of the parameters K_0 and τ_1 can be chosen free. When τ_1 is chosen free (e.g. $\tau_1 = 20 \, \mu s$), we get for K_0

$$K_0 = \frac{\omega_T^2 \, \tau_1}{K_d} = \frac{(0.1 \cdot 2\pi f_S)^2 \tau_1}{K_d} \tag{10.22}$$

with $K_d = 1/\pi$.

The loop filter is realized in most cases as a digital filter, i.e., an infinite impulse response filter (IIR filter). The transfer function $H_{LF}(s)$ must therefore be converted into a discrete time transfer function $H_{LF}(z)$. Using the bilinear z transform [3] $H_{LF}(z)$ becomes

$$H_{LF}(z) = \frac{\left[1 + 2\frac{\tau_2}{T_f}\right] z + \left[1 - 2\frac{\tau_2}{T_f}\right]}{2\frac{\tau_1}{T_f}(z - 1)} \tag{10.23}$$

Here T_f is the sampling interval. The sampling frequency $f_f = 1/T_f$ of the digital filter must be chosen large enough to get sufficient time resolution. It should be a multiple of the symbol rate, i.e., $f_f = OS \cdot f_S$ with OS = oversampling factor. Choosing $OS = 16$ is adequate in most applications; hence, we get

$$T_f = \frac{T}{OS}$$

with T = symbol interval.

The VCO can be realized in the same way by a digital filter. The transfer function of the VCO must also be converted into a discrete time transfer function $H_{VCO}(z)$. Using the impulse invariant z transform [3], we get

$$H_{VCO}(z) = \frac{T_f K_0}{1 - z^{-1}} \qquad (10.24)$$

10.3.3   Frequency Control Loop (Blocks RRCF, Phase Detector, LF, VCO)

The loop that provides locking onto the carrier frequency in both frequency and phase is made up from function blocks RRCF, Phase Detector, LF and VCO (cf. Fig. 10.8). The mathematical model of this control loop is similar to the model used for the conventional Costas loop for BPSK as shown in Fig. 3.1b. The model for the Costas loop for QAM is shown in Fig. 10.16.

Block PD is identical with block "Phase Detector" in Fig. 10.8. The operating principle of the phase detector will be discussed later in Sect. 10.3.4. For the moment, we note that the phase detector is a block having transfer function $H_{LF}(s) = K_d$, with $K_d = 1$. The parameters of the remaining building blocks must be specified such that the control loop is stable. This will be done by means of the Bade diagram. The transfer function $H_{RRCF}(s)$ has been given in Eq. (10.5). For the loop filter, a PI filter is used. Its transfer function $H_{LF}(s)$ known, cf. Eq. (10.18), and the transfer function $H_{VCO}(s)$ of the VCO is given in Eq. (10.19). The parameters τ_1, τ_2, and K_0 must now be specified. The open loop transfer function $G_{OL}(s)$ of the control loop becomes

$$G_{OL}(s) = \frac{K_d K_0}{s} \cdot \frac{1 + s\tau_2}{s\tau_1} \cdot H_{RRCF}(s) \qquad (10.25)$$

The analysis of stability becomes easier when we can find a simplified presentation for the transfer function $H_{RRCF}(s)$. As we will recognize in short, the bandwidth of the RRCF (approximately half the symbol rate f_S) is much larger than the transit frequency ω_T of the open loop gain G_{OL}; hence, it is acceptable to set the magnitude of HRRCF(ω)

$$|H_{RRCF}(\omega)| = 1$$

Fig. 10.16 Mathematical model of the frequency control loop

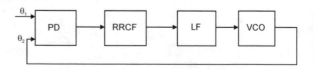

The RRCF can then be replaced by a delay block. We have seen previously that the truncated impulse response of the RRCF has been restricted to the time interval $-nT < t < nT$, with T = symbol interval and n = integer in the order of about 4. Consequently, the RRCF has a delay of $T_d = nT$, and its transfer function can be expressed as

$$H_{RRCF}(\omega) = \exp(-\omega T_d) \tag{10.26}$$

To construct the Bode diagram for $G_{OL}(\omega)$, we first calculate the magnitude $|GOL(\omega)|$. Using Eqs. (10.25) and (10.26), we get

$$|G_{OL}(\omega)| = \left| \frac{K_d K_0}{s} \cdot \frac{1 + s\tau_2}{s\tau_1} \right| \tag{10.27}$$

To get the phase plot of the Bode diagram, we calculate the phase of the open loop transfer function from

$$\phi(\omega) = \phi_1(\omega) + \phi_2(\omega) = phase\left(\frac{K_d K_0}{s} \cdot \frac{1 + s\tau_2}{s\tau_1} \right) - \omega T_d \tag{10.28}$$

In this expression, the phase $\phi(\omega)$ has been split into two components $\phi_1(\omega)$ and $\phi_2(\omega)$ where $\phi_1(\omega)$ is the phase of $\frac{K_d K_0}{s} \cdot \frac{1+s\tau_2}{s\tau_1}$ and $\phi_2(\omega)$ is the phase $-\omega T_d$ of the RRCF. Figure 10.17 shows the magnitude and phase plots of the Bode diagram of the open loop transfer function $G_{OL}(\omega)$. The upper trace is the magnitude plot. Below the corner frequency $\omega_C = 1/\tau_2$ of the loop filter, the magnitude rolls off with -40 dB/decade. Above ω_C the gain rolls off at -20 dB/decade. As shown in the middle trace phase, $\phi_1(\omega)$ is $-180°$ at low frequencies and approaches $-90°$ at high frequencies. At the corner frequency ω_C, ϕ_1 is $-135°$. Note that ϕ_1 is never more positive than $-90°$. The phase $\phi_2(\omega)$ of the RRCF varies linearly with frequency. To get a stable system, the total phase $\phi(\omega)$ must be more positive than $-180°$ at the transit frequency ω_T. This implies that ω_T must be specified such that $\phi_2(\omega)$ is greater than $-90°$. Therefore, we specify the transit frequency ω_T to be that frequency where the phase of the RRCF is $-45°$. Using Eq. (10.26), we get

$$-\omega_T T_d = -\pi/4 \text{ or } \omega_T = \frac{\pi}{4 T_d} \tag{10.29}$$

At $\omega = \omega_T$, phase $\phi_1(\omega)$ must be more positive than $-135°$, otherwise the system would become unstable. When the corner frequency ω_C is chosen $\omega_T/3$, ϕ_1 becomes $-110°$; hence, the total phase ϕ is $-155°$. This results in a phase margin of $25°$, which is sufficient to get a stable system.

Given corner frequency ω_C, the remaining parameters K_0 and τ_1 can be determined. As shown in the magnitude plot of Fig. 10.18, the magnitude of G_{OL} must be 3 at $\omega = \omega_C$. Hence, we have

Fig. 10.17 Bode diagram of
the frequency control loop

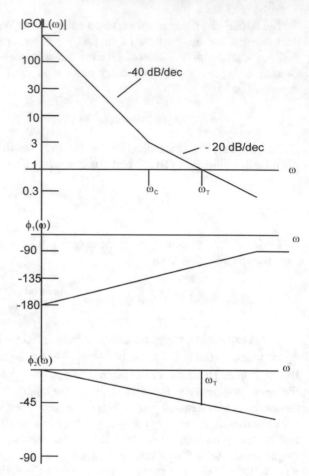

Fig. 10.17 Bode diagram of the frequency control loop

$$3 = \frac{K_0 K_d}{\omega_c^2 \tau_1} \tag{10.30}$$

This is an overdetermined equation for K_0 and τ_1. It is therefore possible to choose τ_1 free and compute K_0 from Eq. (10.30). This yields

$$K_0 = \frac{3\,\omega_c^2\,\tau_1}{K_d} \tag{10.31}$$

The loop filter is realized in most cases as a digital filter, i.e., an infinite impulse response (IIR) filter. The transfer function $H_{LF}(s)$ must therefore be converted into a discrete time transfer function $H_{LF}(z)$. Using the bilinear z transform [3], $H_{LF}(z)$ becomes

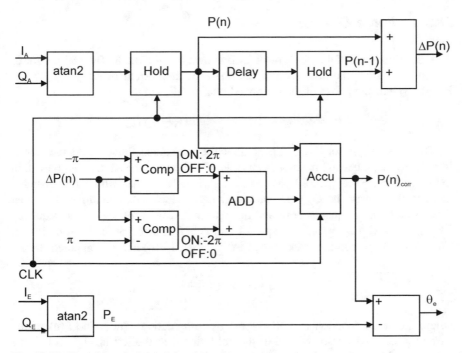

Fig. 10.18 Operating principle of the phase detector

$$H_{LF}(z) = \frac{\left[1 + 2\frac{\tau_2}{T_f}\right]z + \left[1 - 2\frac{\tau_2}{T_f}\right]}{2\frac{\tau_1}{T_f}(z - 1)} \qquad (10.23)$$

Here T_f is the sampling interval. The sampling frequency $f_f = 1/T_f$ of the digital filter must be chosen large enough to get sufficient time resolution. It should be a multiple of the symbol rate, i.e., $f_f = OS \cdot f_S$ with $OS =$ oversampling factor. Choosing $OS = 16$ is adequate in most applications; hence, we get

$$T_f = \frac{T}{OS}$$

with $T =$ symbol interval.

The VCO can be realized in the same way by a digital filter. The transfer function of the VCO must also be converted into a discrete time transfer function $H_{VCO}(z)$. Using the impulse invariant z transform [3], we get

$$H_{VCO}(z) = \frac{T_f K_0}{1 - z^{-1}} \qquad (10.24)$$

10.3.4 Phase Detector

The block diagram of the phase detector is shown in Fig. 10.18. The phase error θ_e is computed from the difference of the phase of the current phasor (I_A, Q_A) (cf. Fig. 10.8) and the phase of the estimated phasor (IE, QE)

$$\theta_e = phase(I_A + jQ_A) - phase(I_E + jQ_E)$$

The phase of phasor (I_A, Q_A) is computed by the atan2 function. As mentioned in Sect. 10.2.3, the atan2 function is capable of computing the arc tg function within the range form $-180°$ ... $180°$, as visualize in Fig. 10.10. As will be demonstrated in short, the phasor (I_A, Q_A) can reach very high values, i.e., a multiple of π when there is an initial frequency error and the phasor (IA, QA) rotates with a frequency $\Delta f = f_C - f_2$ with f_C is carrier frequency and f_2 is frequency of the local oscillator. The phase of the current phasor (IA, QA) is labeled P(n) and is computed from

$$P(n) = arctg\left(\frac{Q_A}{I_A}\right)$$

where the atan2 function is used for arc tg computation. The value P(n) of the currently received phasor is hold in the hold circuit to the left in Fig. 10.18. The content of this hold circuit is updated at the clock rate, which is identical with the symbol rate. One value of P(n) is computed in each symbol period. A second hold circuit, shown at the right in the figure, is used to store the value of the previously received phasor. This value is labeled P(n − 1). Now we use the difference

$$\Delta P(n) = P(n) - P(n-1)$$

to check whether or not the phase P(n) has crossed the 180° boundary, cf. Fig. 10.10. Assume that the previous phase P(n − 1) had a value slightly less than 180° and the current phase crossed the 180° boundary, i.e., is slightly more than 180°, the atan2 function now delivers a negative value in the order of −180°. The difference $\Delta P(n)$ is now applied to the input of a comparator, cf. the upper Comp block in Fig. 10.18. This comparator switches to the logical 1 state when the signal at the input becomes more negative than the value $-\pi$ applied to the + input. The comparator is configured such that its output is 2π in the 1 state and 0 in the logical 0 state. Whenever phase P(n) crosses the 180° boundary, the upper comparator delivers an output signal of 2π. This signal is added to the current content of the accumulator (labeled Accu). The content of the accumulator is updated on every clock signal. The output signal of the accumulator is now added to signal P(n), and the output signal of the accumulator is the corrected phase $P(n)_{corr}$. When the initial frequency error is large, phasor (I_A, Q_A) can execute a number of full revolutions; hence, the corrected value $P(n)_{corr}$ can exceed a multiple of 2π. The phasor (I_A, Q_A) can also rotate in the opposite sense, i.e., clockwise. When P(n) is negative and crosses the 1800 boundary in the opposite direction (i.e., from quadrant Q3 to quadrant Q2), the phase P(n) executes a jump of 2π. Another comparator (bottom in Fig. 10.18) compares $\Delta P(n)$ with π. Whenever $\Delta P(n)$ becomes greater than π, the second comparator generates an output signal of -2π. This value is subtracted from the current content of the accumulator. Phase P(n) can therefore also reach large negative values, i.e., values more negative than a multiple of 2π. Using this measuring method for P(n) yields a much larger pull-in range, compared with the classical methods of phase computation without correcting actions.

As shown in Fig. 10.8, the Estimator (Estim blocks) delivers an estimate (I_E, Q_E) at every clock impulse. The phase of estimated phasor (I_E, Q_E) is again computed by the atan2 function, cf. Fig. 10.18. The phasor error θ_e is now computed by the difference of corrected phase $P(n)_{corr}$ and phase of phasor (I_E, Q_E).

10.3.5 Preamble and Acquisition Process

In Sect. 10.2.2, it was demonstrated that a data transmission must always start with a preamble, because the gain of the AGC amplifier must be set to the correct level. The preamble consists of a series of known symbols. The simplest preamble is a sequence of equal symbols. More frequently a so-called S1 sequence is used [3]. The S1 sequence is made from two symbols S_1 and S_2, as shown in Fig. 10.19a. In

Fig. 10.19 a Phasor presentation of S1 sequence (preamble). The S1 sequence consists of two phasors, S_1 and S_2. **b** Phase P of the S1 sequence with phase error $\theta_e = 0$. **c** Phase P of the S1 sequence with frequency error $\Delta f \neq 0$. The phase error $\theta_e(t)$ is a ramp function. P(0), P(1) ... are the phases of phasor P sampled by the clock signal CLK (cf. Fig. 10.19) at the symbol rate

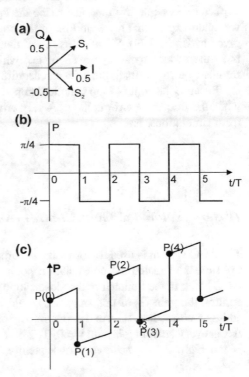

this example, S_1 is a phasor with the coordinates (0.5, 0.5), and S_2 is a phasor with the coordinates (0.5, −0.5). In the S1 sequence, these two symbols are transmitted in alternation, i.e., S_1, S_2, S_1, S_2 ... The phase of S_1 is $\pi/4$, and the phase of S_2 is $-\pi/4$. As shown in Fig. 10.18, the phase P(n) is memorized in a hold circuit at every symbol interval. When no phase error θ_e exists, P(n) is an alternating sequence having the values $\pi/4$ or $-\pi/4$, respectively. This is plotted in Fig. 10.19b. When there is an initial frequency error Δf, the phasor (I_A, Q_A) rotates with frequency Δf; hence, its phase ramps up as shown in Fig. 10.19c. The phase detector must now decide at every clock event whether the current phasor is S_1 or S_2. This is done by building the difference $\Delta P(n) = P(n) - P(n - 1)$ at every clock event. When $\Delta P(n)$ is positive, it is decided that the current symbol is S_1, and when $\Delta P(n)$ is negative, it is decided that the current symbol is S_2. This works as long as the phase rotation within on symbol interval remains below $\pi/2$. This is the case in Fig. 10.19. But when the phase rotation becomes greater, this decision can no longer be made. We conclude therefore that the pull-in range of this type of Costas loop is limited to a value

$$\Delta f_P = f_S/4 \tag{10.32}$$

When there is an initial frequency error, simulations show that the phase error θ_e exhibits a damped oscillation whose frequency is the natural frequency ω_n of the frequency control loop. Because the system is nonlinear, it is difficult to give an explicit expression for ω_n. Considering the open loop frequency response GOL(ω), we could expect that ω_n would be in the order of ω_T (transit frequency), which is given by Eq. (10.29). Simulations reveal that the duration of the pull-in process T_P is in the order of one cycle T_C of ω_n, which is $T_C = 2 \pi/\omega_n$. When the initial frequency error is small, the loop can acquire lock within less than one full cycle of ω_n. Because the control loop is underdamped, T_P can become longer than one cycle when the frequency error is large. A very crude approximation, obtained by results from simulations, is

$$T_P = (0.5...3) \cdot \frac{2\pi}{\omega_n} \tag{10.33}$$

10.3.6 Automatic Gain Control (AGC)

The AGC system is used to normalize the amplitudes of the received symbols I and Q in order to match the constellation points as shown in Fig. 10.9. Consider the AGC block in the I branch of the Costas loop as depicted in Fig. 10.8. Without gain control, the gain G_0 of this amplifier would be 1. It can be modified, however, by adding a correction ΔG. As demonstrated, above block **delivers** a correcting signal ΔA at every sampling instant t = 0, T, 2 T ... When the gain of the AGC amplifier is too high, ΔA is positive. Consequently, the gain of the amplifier should be

Fig. 10.20 Block diagram of
the AGC circuit

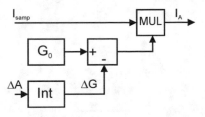

reduced. The block diagram of the AGC amplifier is shown in Fig. 10.21. The amplifier gain is adjusted by applying ΔA to the input of an integrator (labeled Int). The transfer function of the integrator is given by

$$H_{int}(s) = \frac{1}{sT_i} \qquad (10.34)$$

with T_i = integrator time constant. When developing Simulink models for this type of Costas loop, the integrator time constant was chosen by "trial and error." It showed up that a good choice is $T_i = 4\,T$, with T = symbol interval. When the amplifier gain has settled to the required value, ΔA is 0 on average, and the output signal ΔG of the integrator stays almost constant. In most practical circuits, the integrator will be realized by a digital filter, hence the transfer function $H_{int}(s)$ must be converted into a discrete time transfer function $H_{int}(z)$. Using the impulse invariant z transform, $H_{int}(z)$ is given by

$$H_{int}(z) = \frac{T_f}{T_i} \cdot \frac{1}{1 - z^{-1}}$$

T_f is the sampling interval. The sampling frequency f_f is usually a multiple of the symbol rate f_S, i.e., $f_f = OS\,f_S$, with OS = oversampling factor. Hence, we have

$$T_f = \frac{1}{f_f} = \frac{1}{OS \cdot f_S} = \frac{T}{OS}$$

The AGC amplifier in the Q branch is identical with the circuit shown above (Fig. 10.20).

10.4   Simulating the Costas Loop for QAM

A data transmission system for QAM has been simulated by model QAM16_Nyq_mod1C. mdl, Fig. 10.21. The entity of all function blocks in this model is so large that the full block diagram could not be shown in one single figure, but the system is subdivided into a number of subsystems. Subsystem TX

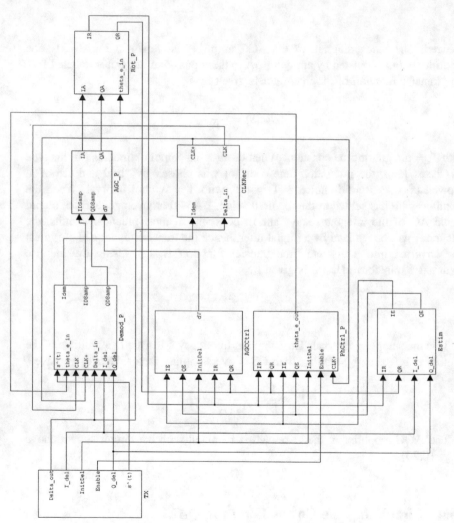

Fig. 10.21 Simulink model QAM16_Nyq_mod1C for QAM data transmission

represents the transmitter. It first generates a preamble, a S1 sequence, followed by a series of random symbols (I, Q). Referring to Fig. 10.8, subsystem Demod_P includes both multipliers MUL, the RRCF filters RRCF1 and RRCF2, and the filters RRCF1 and RRCF2. The clock recovery circuit is in block CLKRec. Subsystem PhCtrl_P contains the circuits of the "Phase Detector," and AGCCtrl contains the circuits of block "AGC Control." Subsystem Estim includes both blocks "Estim" in Fig. 10.8. Subsystem Rot_P is an additional function not yet discussed previously. Its function will be explained below.

All function blocks have been designed using the procedure described in Sect. 10.3. The key parameters of the Costas loop are the following

- Carrier frequency f_C = 400 kHz.
- Symbol rate f_S = 100'000 symbols/s, symbol interval = 10 µs.
- Number of constellation points = 16.
- Duration of impulse response of the RRCF filters is $-nT < t < nT$, with n = 4, i.e., -40 µs $< t < 40$ µs.
- RRCF filter delay $T_d = nT = 40$ µs.
- Transit (radian) frequency ω_T = 19'625 rad s^{-1} (cf. Eq. 10.29).
- The natural frequency ω_n of the Costas loop can be approximated by ω_T, hence on full cycle of ω_n has a duration of $\frac{2\pi}{\omega_n}$ = 320 µs.
  It can be expected that the pull-in time of the loop is in the range 160 … 960 µs, depending on the size of the initial frequency error.

The model is described in detail in the model description included in the Simulink model. To get that description, the following actions are required:

- Start the model QAM16_Nyq_mod1C,
- Click the file menu,
- Click the submenu Model description,
- Click "Description."

Before entering into simulations, the function block Rot_P is explained. This is a phasor rotator, as already seen in previous chapters, cf. Fig. 8.1 [10]. We have seen that the frequency correcting loop has a relatively low bandwidth ω_T (19'625 rad s^{-1}) or $f_T \approx 3$ kHz. If a broadband noise signal is superimposed to the QAM signal as generated by the transmitter, fast perturbations cannot be canceled by the slow frequency control loop. For this reason, a faster correcting loop is added. The block diagram of subsystem Rot_P is displayed in Fig. 10.22. Rot_P contains the blocks LF (loop filter), Int (Integrator), and Rotator, which is the same circuit as the phasor rotator in Fig. 8.1, cf. lower part of the figure. The two atan2 blocks and the adder are part of the phase detector, which is realized in subsystem PhCtrol_P. The phasor (I_A, Q_A) is applied to the input of a phasor rotator that rotates (I_A, Q_A) by an angle ϕ_{rot}. Assume for the moment that $\phi_{rot} = 0$ initially, so the phasor (I_R, Q_R) at the output of the rotator is identical with phasor (I_A, Q_A). When the loop is perfectly locked and there exists no broadband noise, phasor (I_R, Q_R) would be identical with the estimated phasor (I_E, Q_E). Under real conditions,

Fig. 10.22 Block diagram of
the Phasor Rotator (Block
Rot_P)

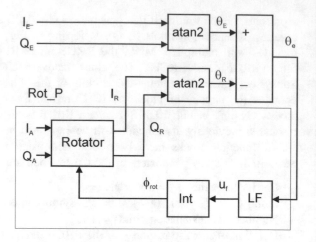

however, there is a deviation, and consequently, the phase detector creates a phase
error signal θ_e that is not zero. Now the loop in block Rot_P comes into action. The
phase error signal θ_e is applied to the input of a loop filter (LF), built from a PI
filter. The output signal u_f of the loop filter drives the input of an integrator, and the
output signal of the integrator is the rotating angle ϕ_{rot}. Consequently, phasor (I_A,
Q_A) is rotated until the rotated phasor (I_R, Q_R) is identical with phasor (I_E, Q_E). The
entire loop in Fig. 10.22 can be considered a PLL having a reference signal θ_E
[phase of phasor (I_E, Q_E)]. The second input of the phase detector (given by the
subtractor on top right) is signal θ_R, the phase of phasor (I_R, Q_R). The remaining
blocks of the PLL are LF and Int. The rotator can be modeled by a simple adder,
because it computes the sum $\theta_A + \phi_{rot}$, where θ_A is the phase of phasor (I_A, Q_A).

The open loop transfer function of this circuit is now given by

$$G_{OL}(s) = \frac{\Theta_R(s)}{\Theta_A(s)}$$

where Θ_R and Θ_A are the Laplace transforms of phases θ_R and θ_A, respectively.
The detector gain $K_d = 1$ in this application, the transfer function of the loop filter
given by

$$H_{LF}(s) = \frac{1 + s\,\tau_2}{s\,\tau_1}$$

with τ_1, τ_2 = time constants, and the transfer function of the integrator is given by

$$H_{int}(s) = \frac{K_0}{s}$$

with K_0 = integrator gain. The open loop transfer function becomes therefore

$$G_{OL}(s) = \frac{K_0 K_d}{s} \cdot \frac{1 + s\,\tau_2}{s\,\tau_1}$$

The parameters K_0, τ_1, and τ_2 are again determined using the Bode plot for the magnitude of $GOL(\omega)$, cf. Fig. 10.15. First the transit frequency of the loop must be chosen. The phasors of the loop are updated at a frequency of 100 kHz; hence, it is reasonable to set f_T at 50 kHz, i.e., $\omega_T = 2\,\pi\,f_T = 314'000$ rad/s. When the radian corner frequency $\omega_C = 1/\tau_2$ of the loop filter is chosen to coincide with ω_T, the phase margin becomes 45°, which is sufficient for stability. This yields $\tau_2 = 3.2$ μs. Finally, the remaining parameters K_0 and τ_1 must be chosen such that the open loop gain becomes 1 at $\omega = \omega_T$. This yields

$$1 = \frac{K_0 K_d}{\omega_T^2\,\tau_1} \quad \text{or} \quad K_0 = \frac{\omega_T^2\,\tau_1}{K_d}$$

As done previously, τ_1 can be chosen free. Specifying $\tau_1 = 10$ μs, we obtain $K_0 = 985'900$.

The loop filter is realized in most cases as a digital filter, i.e., an infinite impulse response (IIR) filter. The transfer function $H_{LF}(s)$ must therefore be converted into a discrete time transfer function $H_{LF}(z)$. Using the bilinear z transform [3] $H_{LF}(z)$ becomes (cf. Eq. 10.23)

$$H_{LF}(z) = \frac{\left[1 + 2\frac{\tau_2}{T_f}\right] z + \left[1 - 2\frac{\tau_2}{T_f}\right]}{2\frac{\tau_1}{T_f}(z - 1)}$$

Here T_f is the sampling interval. The sampling frequency $f_f = 1/T_f$ of the digital filter must be chosen large enough to get sufficient time resolution. It should be a multiple of the symbol rate, i.e., $f_f = OS \cdot f_S$ with OS = oversampling factor. Choosing OS = 16 is adequate in most applications, hence we get (cf. Eq. 10.24)

$$T_f = \frac{T}{OS}$$

with T = symbol interval.

The VCO can be realized in the same way by a digital filter. The transfer function of the VCO must also be converted into a discrete time transfer function $H_{VCO}(z)$. Using the impulse invariant z transform [3], we get

$$H_{VCO}(z) = \frac{T_f K_0}{1 - z^{-1}}$$

Because the inputs of block Rot_P are phasors (IA, QA) and IE, QE)
(cf. Fig. 10.22), this type of loop is also referred to as "phasor locked loop" [10].

10.4.1 Simulations with Model QAM16_Nyq_mod1C

Section 6 of the model description gives a number of hints for simulations with the
model. Here are some suggestions for testing:

Start the model. A parameter window is also displayed where the operator can
enter a number of parameters. Specify number of symbols in data sequence = 20
and number of symbols in preamble = 20, set frequency error = 1000 Hz. Hit the
"Done" button to store that data. Start the simulation. To see whether or net the loop
has acquired lock, watch scope labeled "IE, QE" in subsystem Estim. This scope
shows signals I_{del} and Q_{del}. These are the I, Q signals generated by the transmitter,
but delayed by the accumulated delay of all filters (RRCF in transmitter, RRCF in
Costas loop, etc.). I_E and Q_E are the estimates of I, Q. When the loop has locked, I_E
is identical with I_{del} and Q_E is identical with Q_{del}.

Go back to the parameter window and enter other parameters, e.g., number of
symbols in data sequence = 20 and number of symbols in preamble = 60. Set the
frequency error to 5000 Hz. Watch scope labeled "Scope" in subsystem
PhCtrl_P. Look at the waveform for P(n), the phase of phasor (I_R, Q_R). Because the
actual phase becomes larger than 180°, P(n) exhibits large discontinuities when the
180° boundary is crossed. Now watch signal P(n)corr. This is the corrected phase P
(n). It is a smooth function of time.

Repeat this simulation with a larger frequency error. Enter number of symbols in
data sequence = 20 and number of symbols in preamble = 100. Set frequency
error = 20'000. Start the simulation and watch "Scope" in subsystem PhCtrl_P
again. Look at the waveform for theta_e_pre. This is the phase error computed
during the preamble interval. It shows up that the phase error takes large values, up
to 10 rad (corresponding to 570°).

A lot of detailed information is also available from the callback function
InitFctQAM16_Nyq_mod1C.m. This function is executed whenever a simulation is
started. This function computes all circuit parameters (e.g., time constants, transit
frequency) and sets the parameters of the Simulink function blocks correspondingly
(Table 10.1).

Table 10.1 Pull-in time T_P als a function of initial frequency error Δf

Frequency error Δf (kHz)	Pull-in time T_P (μs)
1	150
5	450
10	750
20	800
25	1000

References

1. U. Rohde, J. Whitacker, *Communications Receivers, DSP, Software Radios, and Design* (McGraw-Hill, New York, 2001)
2. H. Nyquist, Certain topics of telegraph transmission theory. Trans. Am. Inst. Electr. Eng. **47**, 617–644 (1928)
3. R.E. Best, *Phase-locked Loops, Design, Simulation, and Applications*, 6th edn. (McGraw-Hill, New York, 2007)
4. R.J. Higgins, *Digital Signal Prcessing in VLSI* (Prentice Hall, Englewood Cliffs, 1990)
5. A.V. Oppenheim, R.W. Schafer, *Discrete-Time Signal Processing* (Prentice-Hall, Englewood Cliffs, 1989)
6. U. Mengali, *Synchronization Techniques for Digital Recivers* (Springer, New York, 2013)
7. F.M Gardner, A BPSK/QPSK timing-error detector for sampled receivers. IEEE Trans. Commun. **COM-34**(5), 423–429 (1986)
8. F.M. Gardner, *Phase-lock Techniques*, 2d edn. (Wiley, New York, 1979)
9. U.L. Rohde, *Microwave and Wireless Synthesizers, Theory and Design.* (Wiley, 1997)
10. H. Meyr, M. Moeneclaey, S.A. Fechtel, *Digital Communication Receivers*, 2nd edn. (Wiley, 1997)

Index

A

Amplitude error, 125–127
Amplitude modulation (AM), 6–9
    suppressed carrier, 8, 9
Analog Costas loop, 29
Automatic Gain Control (AGC), 125–129,
    143–145, 147

B

Bandpass filter, 130, 135, 136
Binary Phase Shift Keying (BPSK), 1, 8–11,
    14, 29, 31–33, 36, 37, 39, 40, 43, 49–52,
    57–59, 73, 75, 77, 84, 91, 94, 97, 100,
    103–105, 110, 111, 117, 133, 136, 138
Bits per symbol, 77
Bode diagram, 29, 137, 139, 140
Bode plot, 16, 17, 30, 44, 52, 53, 67, 68, 79,
    149
Brickwall filter, 119–122
Butterworth filter, 135

C

Carrier signal, 6, 32, 49
Clock recovery, 128, 129, 131, 134, 147
Complex carrier, 51, 65, 86, 88, 113, 114
Complex envelope, 65, 66, 88, 114
Complex phasor, 86
Complex plane, 51, 67
Constellation diagram (QAM), 117, 118, 126
Cosine carrier, 35, 118

D

Damping factor, 13, 17, 30, 44, 53, 58, 68, 73,
    80, 85
Delay block, 20, 21, 139
Differential encoding, 11

Digital-Controlled Oscillator (DCO), 15, 89,
    93, 99
Digital Costas loop, 31, 32, 43, 45, 46, 58, 59,
    74, 75, 85

E

Estimator, 128, 143
Excess bandwidth, 120, 122
Exor, 96, 100, 131, 136

F

False-locking, 12
Finite Impulse Response Filter (FIR), 49, 119,
    120, 123, 133–135

H

Highpass filter, 6, 8
Hilbert transform, 50, 65, 78, 88, 103, 104, 113
Hilbert transformer, 49, 50, 103

I

In-phase signal, 117
Inter Symbol Interference (ISI), 121, 123, 124
I signal, 32, 35, 91

L

Linear model, 14, 15, 17, 35, 36, 50–53, 67,
    68, 78–80
Local oscillator, 13, 78, 104, 106, 107, 113,
    114, 124, 127, 142
Lock range, 13, 17–19, 21, 28, 30, 37, 45, 53,
    54, 68, 69, 74, 80, 85
Lock time, 13, 17, 18, 30, 37, 45, 47, 53–55,
    59, 60, 68–70, 74, 80, 81, 85
Loop filter, 5, 6, 11, 15, 16, 18, 21, 23, 25, 27,
    31, 36, 37, 39, 40, 45, 52, 54–56, 59, 67,

69–71, 74, 78–80, 82, 83, 86, 88, 89, 99,
    132, 136–140, 147–149
Lowpass filter, 6, 8, 9, 11, 15, 16, 18, 20–22,
    29, 30, 39, 40, 44, 52, 55, 70, 79, 83, 90,
    93, 107, 122, 123
Lowpass filter, ideal, 119

M
M-ary phase Shift keying (mPSK), 1, 78, 79,
    82, 84, 86, 87
Matlab, 2, 3, 32, 123, 127, 135, 136
Modified Costas loop, 1, 49, 50, 52, 56, 57, 59,
    61, 65–67, 71–73, 77–79, 82–84, 90,
    103–105, 113, 114
    for mPSK, 78, 79, 82
    phasor rotator, 1, 103–105, 113, 114
        for QPSK, 1, 113, 114
Modulation index, 6

N
Natural frequency, 13, 17, 18, 30, 44, 53, 58,
    68, 73, 80, 85, 144, 147
Non linear model, 19, 20, 25, 70–72, 39, 40,
    42, 55, 56, 82, 83
Nyquist filtering, 123, 124

O
Octant, 108, 109
Oversampling factor, 89, 96, 98, 99, 104, 109,
    110, 114, 133, 135, 137, 141, 145, 149

P
Phase ambiguity, 90
Phase detector, 5, 11, 13–15, 18, 25, 29, 36, 37,
    40, 43, 52, 59, 60, 62, 66, 78, 86, 88,
    127, 131, 132, 136, 138, 141, 142, 144,
    147, 148
    gain, 11
    with extended phase error range, 25, 60
Phase error, 5, 6, 9, 11, 15, 25, 35–37, 51, 52,
    60, 62, 66, 78, 86, 88, 98, 125, 127, 128,
    142–144, 148, 150
Phase error band, 98
Phase/frequency detector, 5, 11, 13–15, 18, 25,
    29, 36, 37, 40, 43, 52, 55, 60, 62, 71, 83
Phase-Locked Loop (PLL), 5–9, 13, 131, 132,
    136, 148
Phase step, 94, 98, 99, 104, 110, 114
Phase transfer function, 13, 132
Phasor, 51, 52, 54, 60, 62, 66, 67, 69, 80, 86,
    90, 93, 94, 96, 98, 100, 103, 104, 108,
    109, 113, 125–128, 142–144, 147–150
Phasor control, 104, 109, 114

Phasor rotator, 1, 94–100, 103–105, 108, 110,
    111, 113, 114, 147, 148
Preamble, 1, 12, 59–62, 91, 126–128, 143, 147,
    150
Pre-envelope signal, 49–51, 65, 77, 86, 88,
    103, 104, 113, 114
Prefilter algorithm, 129
Prefilter method (Gardner), 129
Prewarping, 31, 45, 59, 74, 85, 86, 100, 111
Pull-in range, 13, 19, 25–27, 31, 34, 40, 42,
    45–47, 55, 57, 71, 72, 83, 84, 90, 98, 99,
    110, 114, 143
Pull-in time, 13, 19, 24–28, 32, 40, 42, 43, 46,
    47, 57, 59, 72, 75, 84, 90, 98, 147, 150

Q
Q signal, 35
Quadrant, 66, 94–96, 127, 142, 143
Quadrature, 89
Quadrature Amplitude Modulation (QAM), 1,
    117, 122–125, 133, 138, 145–147
Quadrature Phase Shift Keying (QPSK), 1, 35,
    39, 41, 43, 46, 65–67, 71–73, 75, 77, 84,
    90, 107–111, 113, 114, 117, 124, 133,
    136
Quadrature signal, 117

R
Raised Cosine Filter (RCF), 122–124,
    128–130, 133–135
Root Raised Cosine Filter (RRCF), 124, 129,
    133–135, 138, 139, 147, 150
Rotating switch, 94, 104, 108, 110
Rotator control, 94, 95, 104, 108

S
S1 sequence (preamble), 143, 147
Simulink, 1–3, 32, 33, 46, 59, 60, 74, 75, 86,
    87, 97, 100, 101, 104–106, 112, 115,
    130, 145–147, 150
Sinc function, 119
Sine carrier, 35, 43, 73, 84, 118
Symbol rate, 20, 29, 30, 32, 43, 44, 57, 73, 84,
    89, 96, 98, 99, 104, 109, 110, 114,
    117–120, 122, 124, 126, 129, 130,
    135–138, 141–143, 145, 147, 149

T
Transfer function (of Hilbert transformer), 49

U
Up/down counter (bidirectional counter), 96

V
VCO gain, 13, 16, 26, 133, 136
Voltage-Controlled Ocsillator (VCO), 5, 6, 8,
      9, 11–13, 15, 16, 18, 20, 21, 23, 25, 26,
      32, 36, 37, 40, 46, 51, 52, 54, 55, 57, 59,
      65, 67, 69–72, 74, 78–80, 82, 83, 86, 93,

99, 114, 133, 124, 127, 132, 133, 136,
138, 141, 149

Z
Z transform, bilinear, 31, 45, 58, 59, 74, 85, 86,
      89, 99, 100, 111, 137, 140, 149

Printed in the United States
By Bookmasters